Choosing the Future

Choosing the Future

Technology and Opportunity in Communities

KAREN MOSSBERGER, CAROLINE J. TOLBERT,
AND SCOTT J. LACOMBE

OXFORD
UNIVERSITY PRESS

Oxford University Press is a department of the University of Oxford. It furthers
the University's objective of excellence in research, scholarship, and education
by publishing worldwide. Oxford is a registered trade mark of Oxford University
Press in the UK and certain other countries.

Published in the United States of America by Oxford University Press
198 Madison Avenue, New York, NY 10016, United States of America.

Library of Congress Control Number: 2021013440
ISBN 978–0–19–758576–4 (pbk.)
ISBN 978–0–19–758575–7 (hbk.)

DOI: 10.1093/oso/9780197585757.001.0001

1 3 5 7 9 8 6 4 2

Paperback printed by Marquis, Canada
Hardback printed by Bridgeport National Bindery, Inc., United States of America

Contents

Tables and Figures

Tables

      Low-Income Broadband Subscription Rates   163

6.2. Inequality in Broadband Subscriptions by Income, Counties   164

6.3. Digital Inclusion Trailblazer Cities, 2020   180

6.4. Predicting April 2020 Unemployment across US Counties   186

      Appendix A Table. Broadband Questions, Current Population
      Survey/American Community Survey   189

Figures

Preface and Acknowledgments

Writing a book about technology, opportunity, and place summons for us our own past experiences and a recognition that we are all shaped by the communities where we have lived, even as we move about and encounter new perspectives.

Collectively, these places have influenced our views about economic opportunity and how technology might play a role. Writing about the trials and the hopes of Detroit was especially poignant for someone who grew up there and worked for city government for more than a decade and a half. St. Louis, like many other heartland cities, has experienced similar challenges, with steep population declines and marked inequality across neighborhoods. In contrast, one of us is originally from Boulder, Colorado, and has returned over the years to watch its transformation through the concentration of the technology industry. Later collaborating with local governments, nonprofits, and community organizations in northeast Ohio and Chicago, we learned about people organizing in different ways to use technology for change in their own cities. Beyond these cities, we have come to know rural communities too. In Iowa, we have seen highly connected university towns in contrast with the not-so-distant rural areas that often remain poorly served. The challenges are different for providing broadband in the rugged, mountainous terrain in other states where we have lived, like Arizona and Colorado. And for Tribal Nations, their history and struggles for Tribal sovereignty are an inherent part of the story of broadband today. The local vision for broadband and the needs for the future will vary across these places.

The evidence we show throughout this book indicates, however, that widespread broadband use and digital information have delivered important economic benefits in all types of communities over the past two decades. Our contribution is to examine the impact of *inclusive broadband,* or the percentage of the local population with broadband subscriptions—a more reliable measure of community use than the data on infrastructure used in most previous studies. We argue that broadband subscriptions and uses represent a form of digital human capital in communities, just as education contributes human capital for economic development.

With new longitudinal data for counties and metropolitan areas estimated for 2000–2017, we demonstrate in these pages the role that broadband adoption has played in how communities have grown and diverged over nearly two decades. We show the economic benefits of broadband connectivity and digital human capital across counties and metros, borne out across multiple models, and how broadband subscriptions in a state are related to policy innovation.

But broadband adoption is not universal across places, and the gaps are not only between rural and urban areas. Recently available census data allows us to map the disparities in broadband subscriptions across neighborhoods within cities. This reveals high dependence on cell phones to go online in some city zip codes, and low rates of internet use overall in others. As COVID-19 showed, mobile-dependent households were not equipped for remote learning, telehealth, telecommuting, and other needs. The consequences of these broadband inequalities were evident in urban and rural communities alike. Just as economist Raj Chetty and his colleagues have demonstrated that place-based inequality has consequences for lifelong economic mobility, we show that spatially patterned disparities in broadband use affect economic opportunity too. As a society, we now face choices that will determine future growth and innovation, and how equitably broadband use and its benefits are distributed within and across communities.

This book represents several years of labor, and we have multiple sponsors and colleagues to thank for their support along the way. A National Science Foundation (NSF) grant via Broadband Community and Capacity (NSF #1338471) funded construction of the book's core data on metros, counties, and states, which we have made publicly available in the Iowa–Arizona State University (ASU) Broadband Data Portal. Through this grant, we used multilevel modeling and data from the Current Population Survey to estimate broadband subscriptions across these geographies between 2000 and 2012, before the US Census Bureau began to collect broadband subscription data in 2013 through the larger American Community Survey.

In the chapter on cities, we discuss the impact of Chicago's Smart Communities digital inclusion program. The evaluation research we report was supported by the John D. and Catherine T. MacArthur Foundation, with citywide surveys that were funded by the Partnership for a Connected Illinois through the Broadband Technologies Opportunity Program. The Institute for Policy and Civic Engagement at the University of Illinois at Chicago also contributed support for that evaluation. The Chicago research was foundational for this book in other ways. Our use of multilevel modeling for neighborhood-level data from the Chicago surveys eventually led us to the idea of estimating data and examining outcomes at different geographic scales.

We would like to thank the Economic Innovation Group and the Brookings Institution for their permission to use their data (the Distressed Communities Index and the Metro Monitor, respectively). These outcome variables allowed us to demonstrate broad impacts in communities. We also appreciate discussions with Rafi Goldberg from the National Telecommunications and Information Administration about the Current Population Survey data.

We have many colleagues to thank for their assistance. Our co-investigators on the NSF grant were Colin Gordon and Julianna Pacheco of the University of Iowa, and William Lehr of the Massachusetts Institute of Technology. Colin brought the data in the portal to life with maps and graphs, Juliana contributed her experience with multilevel models and health policy, and Bill provided his in-depth knowledge of broadband data and research.

In February 2015 we held a workshop at ASU to discuss plans for the portal and uses of the data. We are indebted to these colleagues and students for their expertise and assistance: Marisa Duarte, William Franko, Jon Gant, Ramon Gil-Garcia, Andrea Kavanaugh, Eszter Hargittai, J. Scott McDonald, Sharon Strover, Christine B. Williams, Traci Morris, Erik Johnston, and Brian Gerber. Thanks to Ellen Helsper, Fabricio Senne, Matthew Bui, and Hernan Galperin for sharing knowledge on their projects on broadband, place, and inequality. We have benefited from comments made during our presentations of the data portal and later the findings as we developed the book. Along the way, we engaged with both academics and policymakers at meetings: the Broadband Opportunity Council; Schools, Hospitals and Libraries Broadband Coalition; National Digital Inclusion Alliance; International City-County Management Association; National Association of County Governments; NIC.br (Brazil); Partnership for Progress on the Digital Divide; dg.0 International Research Conference on Digital Government; Association of Internet Researchers; Telecommunications Policy Research Conference; International Political Science Association; American Political Science Association; State Politics and Policy Conference; Texas Tech; George Washington University; and University of Sao Paulo. The comments from the two anonymous colleagues who reviewed our manuscript have strengthened the manuscript substantially.

Many former and current graduate students have contributed to this project in various ways: doing research, organizing workshops, building portals, creating maps, checking references, and coauthoring conference papers. We acknowledge special contributions with coauthorship of specific chapters in the book. More broadly, however, we owe a debt of gratitude to Zhang Yang, who was the primary graduate research assistant working at Iowa on the NSF grant creating the time series broadband subscription data for counties and metros. Professor Yang is now on the faculty at Southwest Jiaotong University, China. We also thank Kellen Gracey, Bomi Lee, and Christopher Anderson from the University of Iowa and Kuang-Ting Tai, Joshua Uelbeherr, Nicholet Deschine, Mattia Caldarulo, and Meredith McCullough from ASU. Senior editor Angela Chnapko has worked with us on more than one project now, and we are grateful for the guidance and support we have received from her and the other professionals at Oxford.

We especially extend our thanks to our families for their patience and support, as well as to our colleagues on other projects who have sometimes wondered when this book would be finished. By tracking outcomes for broadband use across two decades and many geographies, we hope that this contributes a clearer view of digital human capital across communities in the recent past and the solutions so urgently needed for the digital future.

1

Innovation and Inequality

Two Narratives of Place

In the spring of 2018, in a converted factory that now houses a maker space, the Flint Police Department announced a policy challenge with a $30,000 prize for the best idea to use gigabit broadband to solve urgent public safety issues: gun violence and opioid deaths in the city (Ahmad 2018). Flint's challenge was to create new applications for gigabit broadband, which is at least six times faster than the average internet speeds available in the United States today (OECD 2019).[1]

Flint, Michigan, is one of dozens of Smart Gigabit Communities that have joined US Ignite, a partnership launched by the National Science Foundation and the White House Office of Science and Technology Policy. The partnership encourages deployment of fast fiber-optic broadband networks and the development of new applications in participating communities. According to Flint Ignite, gigabit networks will enable residents to build "a newly connected and technologically thriving city: a Flint for the 21st century" (US Ignite n.d.).

This vision, however, is not yet a reality, for Flint has one of the lowest rates of home broadband adoption among the nation's cities; only 59% of households had any type of broadband subscription in 2017, the first year that census data on broadband was available for all geographies in the United States. This 59% of Flint households included 10.4% who had internet service only through their cell phones, and compared with 83.5% with a broadband subscription in the United States (American Community Survey, 2017a, one-year estimates).

Innovations in smart agriculture and precision agriculture are also increasing aspirations for broadband use in rural communities. "Drones have become an important tool," according to an article citing officials from Rosebud County, Montana. "You can take drone flights of your fields, monitor irrigation patterns, you can send a picture of a weed you don't know about to Monsanto or Dupont and they can get back to you with what chemical you

Choosing the Future. Karen Mossberger, Caroline J. Tolbert, and Scott J. LaCombe, Oxford University Press. © Oxford University Press 2021. DOI: 10.1093/oso/9780197585757.003.0001

need to treat it," explained one county commissioner (Ban 2018). Ranches depend on connectivity for many other purposes, including blood sample data to check livestock feed rations and remote repairs for combines run by computers (Ban 2018). Yet just under 42% of Rosebud County had broadband subscriptions in 2017 (American Community Survey [ACS] five-year estimates). With about 9,000 residents, Rosebud County includes a large - part of the Northern Cheyenne Nation's lands (Rosebud County, n.d.a) and is nearly 40% Indigenous.[2] Added to the lack of broadband adoption is slow and unreliable internet service in the area. The Southeastern Montana Development Corporation worries about losing businesses and residents from the region, which has traditionally depended on agriculture, tourism, railroads, and coal mining (Rosebud County, n.d.b). Counties served by the economic development corporation have identified broadband as their most important need (Ban 2018).

Flint and Rosebud County illustrate the long-standing challenge of digital inequality that escalated to crisis proportions during the pandemic of 2020. In a survey conducted by the Pew Research Center in April 2020, 43% of low-income households said their children would have to use cellphones to complete their school assignments, and 40% required public Wi-Fi because they lacked reliable internet access at home (Vogels et al. 2020). As a result, schools and local governments scrambled to find and distribute emergency tablets and to provide hotspots on school buses, in parking lots, in homeless shelters, and outside shuttered libraries (Stewart 2020; Romm 2020; Associated Press 2020). Even these stopgap solutions were inadequate for many students; congregating around a hotspot doesn't replace home connectivity (Bekiempis 2020). As jobless numbers surged and the need for social services increased, nonprofits and government offices went remote and struggled to serve the unemployed and other low-income clients who were not online, adding to the misery and economic suffering created by the pandemic (Goldstein 2020; Quaintance 2020; Johnston 2020a). Local government meetings became entirely virtual (Johnston 2020b), excluding the voices of some residents and neighborhoods. There were public health consequences as well, as residents without regular internet access were at more risk, unable to order food deliveries or to access health care through telemedicine (Bekiempis 2020).

For some communities, this lack of connectivity was unexpected. "It's 2020. Why is the digital divide still with us?" wondered one state and local government magazine in its headline (Simama 2020). Those who have studied or addressed this issue over the years were less surprised. The pandemic did not create but merely exposed deep fault lines in American society, including digital inequalities, that exacerbate racial and economic disparities, and that are patterned by place, affecting communities as well as individuals.

Current federal policies define the issue as a lack of rural infrastructure, but the pandemic has demonstrated that this is an erroneous oversimplification of the problem. In Flint, as in other low-income urban communities, poverty and the lack of affordability are barriers to technology use rather than the lack of broadband infrastructure that exists in some rural areas. Other programs, such as US Ignite, have focused on gigabit speeds as a key to economic revitalization. While superfast internet speeds can support innovative uses, they will not close Flint's broadband deficit for ordinary residents unless there are affordable options to promote widespread adoption. Addressing this dilemma is in fact a priority for the local government, businesses, and other community leaders who are promoting Flint Ignite. They view gigabit technology as a solution for "cost-effectively connecting even the poorest areas of the city with high-speed, reliable internet access, helping to close the increasingly harmful information gap that negatively impacts vulnerable communities" (US Ignite n.d.).

Far from the empty factories and warehouses of Flint and the forests of Rosebud County, Silicon Valley embodies innovation in the information and data-driven digital economy; information technology (IT) is part of the DNA of this region. Unsurprisingly, some of the highest rates of broadband adoption in the United States are in major cities of the area. Sunnyvale, California, ranked among the most digitally connected cities in the nation with almost 93% of households with broadband subscriptions (American Community Survey, 2017, one-year estimates, Table S2801). Sixty-four percent of Sunnyvale households with annual incomes below $20,000 have broadband (wireless or wired), compared to only 42% in Flint; thus, a majority of the poor are connected in Silicon Valley. With 153,000 residents and 23 square miles, this former orchard now has 229 firms related to IT, according to one business listing (www.manta.com), and surrounding tech firms are snapping up property. The *San Francisco Chronicle* described an "arms race" (Lee 2018) for office space in Sunnyvale, with tech giants like Google and Apple spilling over from their headquarters in nearby communities, where the density of Tesla owners is higher than in most other regions of the nation (Brown 2019). Growing Sunnyvale illustrates the tendency of technology firms to cluster together in what economist Enrico Moretti (2012, 15) has called ecosystems of innovation.

Part of this ecosystem is a highly connected populace, in contrast with Flint and Rosebud, where limited opportunity is evident in a lack of digital connections as well as vacant properties or declining populations. Flint and Sunnyvale also represent polar opposites in the economic fortunes of communities today; annual median household income in Sunnyvale, at $134,234, was more than five times Flint's (at $26,330) (American Community Survey, 2017, one-year estimates, Table S1901). For a benchmark, the annual household median income in 2018 was just over $60,000.

In the following pages, we ask how the economic fortunes of communities, urban and rural, are tied to digital skills and technology use in those communities. Broadband use in the general population offers a broader view of technology's impact than the attraction of technology firms, venture capital investment, or the share of IT employment. Broadband subscriptions are a measure of how inclusive technology use is in a community. Prior studies have relied mostly on the presence of broadband infrastructure or providers, but places like Flint demonstrate that use depends on more than whether broadband is available for purchase.

Prior to this study, there was a lack of precise, comparative, and systematic data on broadband subscriptions for communities over time at the subnational level. More than the broadband maps of recent years that show the number of providers, speeds, or the reported availability of service, broadband subscriptions represent use in the population and potential benefits for individuals and their communities.

Broadband, or high-speed internet, "has become vital to almost every segment of the Nation's social and economic fabric," according to the National Broadband Research Agenda (NTIA and NSF 2017, 3), and measures of use over time capture a fuller picture of technology's role in the life of communities. The results we describe in subsequent chapters indicate that addressing inequality in technology use and information access may be more important for closing place-based gaps in the knowledge economy and for creating economic opportunity than previously recognized.

While there is a wealth of publicly available data to measure broadband adoption for individuals over time from the Pew Research Center and government sources, there has not been accurate data to measure broadband subscriptions for substate communities. The US Census Bureau released the first estimates of broadband subscriptions for census tracts in December 2018 (the 2017 five-year estimates), with broadband defined as the percentage of the population with a broadband internet subscription, such as cable, fiber optic, or DSL; mobile/cell phone; or satellite internet service. Since 2013, some community-level broadband data has been available from the US Census Bureau, but only for places with a population greater than 65,000, omitting two-thirds of counties and all other smaller geographies such as neighborhoods in urban areas. Data on broadband subscriptions from the Federal Communications Commission (FCC) has been too imprecise to assess change or differences across communities, reporting subscriptions only in quintiles.

This book fills critical gaps in the government data but also, more importantly, traces the impact of broadband use in communities over time, examining patterns and outcomes for neighborhoods, cities, metros, counties, and states over nearly two decades.

Trends in Innovation and Inequality

Together Rosebud County, Flint, and Sunnyvale symbolize two contemporary place-based narratives about technology and its geographic impacts in the United States: (1) the promise of innovation and (2) the rise of inequality. Places and people left behind were increasingly part of the conversation in politics, policy circles, and the press even before the pandemic. Think tanks have referred to these trends as the "great reshuffling" (Economic Innovation Group 2018, 1) after the 2008 recession, from which cities in the heartland and rural communities, in particular, did not fully recover. Along with the formation of what has been called "superstar cities" (Badger 2018; Berube and Murray 2018; Florida 2017) along the coasts and in other favored tech hubs, there has been a "great divergence" in fortunes between urban regions (Moretti 2012, 73). This "winner-take-all urbanism" (Florida 2017, 6) includes deep disparities within metropolitan areas and city neighborhoods as well as between cities.

This historic change in opportunity across places is in part a result of structural changes in the economy in which technology has played a prominent role, creating new demands for skills in the workforce (Giannone 2017; Moretti 2012, 105). Trends in earnings and income since the 1980s show that there is greater income inequality (Rose 2018; Bartels 2016), a declining share of the economy going to worker pay (Leonhardt and Serkez 2020), and a decrease in labor force participation by less educated prime-age individuals (Krause and Sawhill 2018). Upward mobility has declined as well in recent decades. Only a little over half of the age cohort born around 1980 earn more than their parents did, in comparison with nearly 90% of children born in the 1940s (Chetty and Hendren 2017). As Chetty and colleagues have argued, such developments strike at the heart of the American dream.

Faltering opportunity for many middle- and lower-income Americans, especially those without a college degree, has been accompanied by a loss of faith in the future (Krause and Sawhill 2018). Political and ideological debates about the shrinking middle class have emerged since the 2008 recession, but changes in the economy have been building for decades, with an especially heavy toll on people of color. Trends prior to COVID-19 have set the stage for disparate impacts as the nation grapples with what has been called one of the most challenging problems in American history (Kettl 2020), including a labor market collapse of unprecedented speed, impacting low-wage workers most (Bartik et al. 2020b).

The Geography of Inequality

Disadvantage is also increasingly sorted by place. Comparing metropolitan regions across the United States, Moretti concludes that "your salary depends more on where you live than on your resume" (Moretti 2012, 88). Despite seemingly placeless interactions facilitated by the internet, geography is gaining new significance in the digital era, as the gap between prosperous and distressed communities widens. Regional inequalities have affected formerly industrial areas as well as rural towns in the South and Midwest, feeding divisive and contentious politics, evident in the 2016 election and its aftermath (Sawhill 2018). In the decades following World War II, income differences across communities narrowed, but since the 1980s, skill-biased technological change has instead widened these gaps (Giannone 2017). Places with highly educated populations are flourishing while rural areas and cities with less educated populations founder (Glaeser 2011, 57). According to Moretti, this geographic divergence is deepening and accelerating because of the clustering or agglomeration effects of technology (Moretti 2012, 101); technology companies reside in geographic areas with a dense pool of highly skilled workers.

Current place-based disparities have consequences far into the future. From the metro to the neighborhood level, in rural and urban America, where an individual grows up affects lifelong economic opportunity (Chetty et al. 2018). Research by Chetty and Hendren (2017) finds that location has huge consequences for poverty and intergenerational mobility in the United

States. Children raised in some counties go on to earn much more than they would if they grew up elsewhere, controlling for other factors; the results show that places have a causal influence on upward mobility. Manhattan, New York, for example, ranked low for income mobility for children in poor families (bottom quintile of income distribution), better than only about 7% of counties. In contrast, geographically proximate Bergen County, New Jersey, ranked high for such mobility (better than 88% of counties). The quality of public education and two-parent households are some of the key factors in economic mobility across generations, but the authors do not consider broadband access or use.

While technology has fueled rising inequality through changes in the labor force and spatial concentration of the IT industry, in other ways, it might lead to more inclusive innovation and growth. The National Broadband Plan of 2010 established goals for broadband connectivity because of its anticipated benefits for economic growth and community development across a diversity of communities, as well as other collective benefits. As the Flint scenario indicates, technology innovation is viewed by local leaders there too as a potential solution for years of disinvestment and disadvantage. The smart cities movement is motivated in part by the hope that technology applications will catalyze new local economic development through innovative uses of sensors, artificial intelligence, data analytics, and more. An official report issued by the Smart Cities Council, entitled "Smart Infrastructure Unlocks Equity and Prosperity in Our Cities and Towns," addresses this aspiration (Smart Cities Council 2016). Not every city will become a tech hub, however, and the application of smart city solutions alone will not likely unleash prosperity. But widespread technology use in the community may offer other prospects for human capital development and the future, prospects that lead to greater opportunity throughout a community as well as innovation.

New Data on Broadband Subscriptions over Time

This book provides unique, previously unavailable longitudinal data on broadband subscriptions for local communities over the past two decades, a transformative period in American history. From 2000 to 2018, the percentage of the population with a broadband subscription rose from just 5% to 83% (including mobile and fixed broadband). Despite the lofty goals forged in the National Broadband Plan a decade ago, however, adoption has flattened out and broadband subscriptions have experienced declines in recent years, according to our community-level data and national sources (Pew Research Center 2019; NTIA Data Central 2020).

Nearly 17% of households, or over 56 million people, lack personal access to online information in the form of a broadband subscription (mobile or fixed wireline connection) (American Community Survey, 2017, one-year estimates). They lack the ability to participate in society online, which we have elsewhere called digital citizenship (Mossberger, Tolbert, and McNeal 2008), without services that are increasingly critical for finding jobs, maintaining health, reading the news, completing homework, getting to work, doing work for a job, receiving national weather alerts, etc.

Additionally, of the 83% who have broadband subscriptions, it is estimated that 17% of Americans in 2019 went online using a smartphone, without high-speed home internet (Anderson 2019). This percentage rises to 26% for those with annual household incomes below $30,000 (Anderson 2019). But this national data tells us little about how broadband use is distributed across places.

Researchers have long emphasized the need for more granular, local data on broadband use as a more meaningful measure than deployment (NTIA and NSF 2017, 3; Gillett et al. 2006; Holt and Jamison 2009). Existing data on broadband availability (infrastructure) suffers from many limitations,[3] but precise data on broadband adoption has been especially difficult to obtain at the subnational level. The most granular data, from the FCC, aggregates fixed wireline subscriptions or mobile subscriptions by quintile (0–20%, 20–40%, 40–60%, 60–80%, etc.) for the nation's census tracts. Clearly, there are substantial differences between communities with 40% broadband adoption and 59%, but under the FCC categories both communities would count the same. The categorical data is also less sensitive to change over time, though it has been used in some studies. Because biases in tract-level data can multiply as geographies are scaled up, we rely instead on interval-level data on broadband subscriptions (cell phone or home broadband) (Figures 1.1 and 1.2) derived from modeling and from the US census.

Much like the human genome project, we map broadband adoption, or subscriptions, across the nation's local communities, the component parts of the society. As with the genome project, however, mapping is only a first step. This data allows us to go further, to measure the impact of broadband across communities on policy outcomes.

First, we offer new evidence on disparities in broadband subscriptions for communities defined at multiple scales—for states, counties, metros, cities,

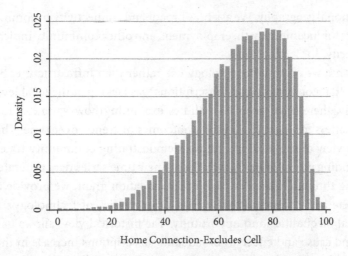

Figure 1.1 Distribution, broadband subscriptions without cell phone only, for 72,000 census tracts (2017 American Community Survey five-year estimates).

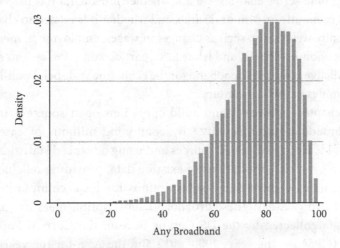

Figure 1.2 Distribution, broadband subscriptions of any type, with cell phones, for 72,000 census tracts (2017 American Community Survey five-year estimates).

and neighborhoods. In contrast with prior studies that only describe national trends, we portray differences across both urban and rural communities, and diverging fortunes from neighborhoods to states. This allows us to connect broadband use with economic outcomes across different layers of the national geography. We ask how broadband connectivity in communities matters for income growth, employment, and other community indicators of prosperity.

Second, we examine technology use rather than infrastructure. Because reliable data on broadband subscriptions was lacking at the local level until recently, there have been few studies examining how variation in adoption matters for local community outcomes or trends over time. Third, we offer a view of the past and future, demonstrating community trajectories and lending rigor to our analysis. Using new time series data generated with funding through a National Science Foundation grant, we provide a more complete and compelling view than prior studies of technology's role in regional inequalities and opportunity. The historical view allows us to understand cause and effect: as broadband subscriptions increase in the community, does this lead to economic growth, controlling for other known predictors? Or are broadband subscriptions merely more prevalent in communities that are already more affluent?

Using time series analysis, we ask whether broadband has played a role in local economic growth or decline, and whether it is related to changes in community well-being, such as changes in wages, employment, median income, economic distress, and labor force participation. We seek to evaluate the predictive power of broadband for the economic and social well-being of communities in the 21st century.

The chapters showcase, and build upon, new open-source data measuring broadband connectivity.[4] By combining millions of cases from households from two public sources and using advanced statistical modeling, we overcome limitations of existing data, providing reliable aggregate estimates of broadband subscriptions for local communities over the past two decades. These broadband subscription measures are built from data collected by the US Census Bureau: the Current Population Survey (CPS) for the years 2000–2012 and the ACS for the years 2013–2017. We rely on the CPS for much of the historical data, since the government has been collecting community-level data on internet subscriptions only since 2013 for its 3-million-person ACS. Yet the CPS data from 2000 was previously available only at the national, state, or individual levels, not for counties, metros, or cities. Multilevel statistical models and poststratification weights are used to create robust comparable estimates of broadband subscriptions for counties and metros from 2000 to 2017 (see also Mossberger et al. 2013).

As mentioned earlier, until the 2017 ACS (released in December 2018), data on broadband subscriptions was available only for geographies with populations of 65,000 or more, so population connectivity went unmeasured in most local US communities.[5] Five-year estimates covering 2013–2017 were made available for the first time for the nation's 72,000 census tracts, but this 2017 data is only for a single year.

We use these new annual measures of broadband subscriptions for communities in time series analysis (2000–2017). This provides important advantages in making causal inferences, as the change in the predictor variable (technology use) precedes the change in the outcome with the use of lagged variables (Angrist and Pischke 2009; Box-Steffensmeier and Jones 2004). Data over time also allow the researcher to determine whether changes in economic outcomes are temporary or persistent, and whether the change in the outcome variable is a response to short-run or long-run trends in broadband adoption in the population. For these reasons, time series models are often considered the gold standard for predictive social science research using observational data and causal inferences and allowing measurement of average treatment effects.

Before turning to our analysis, we consider in the next section why broadband adoption might matter for these economic outcomes, based on theory and prior research grounded primarily in broadband infrastructure deployment.

Innovation and Inequality in the "Human Capital Century"

Technology and globalization have restructured nation-states, further concentrating economic activity in cities, with their information-based economies (Barber 2013). IT emphasizes dense clustering of IT and other knowledge-intensive firms because it "has increased the returns to the knowledge that is best produced by people in close proximity to other people" (Glaeser 2011, 6). An economy based on flows of ideas and information produces more clustering or agglomeration than in the past (Glaeser 2011, 6). The interactions and knowledge spillovers across firms within an area boost productivity and innovation. So, while technology makes interactions possible across long distances, the pull of shared labor markets with talented workers has focused venture capital and start-ups in a smaller number of places. For these "brain hubs" (Moretti 2012, 96) of highly educated workers, the returns are substantial, with large multiplier effects that increase local wages across sectors from food service workers to production.

The returns to agglomeration, or the clustering of firms, boost innovation and productivity but also drive greater inequality across places. Local communities have addressed these new realities by trying to attract technology firms and to foster technology-driven innovation. In the heated competition for Amazon's second headquarters, cities across the country offered inducements reminiscent of the smokestack chasing of earlier eras.

Luring tech firms based on old-economy incentives ignores the dynamics that provide the basis for innovation. Fundamental changes in the economy mean that individual and community fortunes are increasingly driven by the education and skills of the population, by human capital development, in what Moretti has called the "human capital century" (Moretti 2012, 215). Tech hubs cluster economic activity and multiply its impact through human capital externalities that spill over from individuals to society. These human capital externalities are "at the heart of modern economic growth theory" (Moretti 2012, 99; cf. Lucas 1988; Becker 1964).

While local leaders often focus on innovation in terms of the presence of technology firms, network speeds, or the growing Internet of Things, the future of local innovation relies on more than interconnected devices and sensors or the share of IT employment. It also depends on connected networks of residents, businesses, and organizations that have the widespread capacity to *use* technology. This is a form of human capital as well— the ability to employ technology for innovation throughout the myriad interactions that constitute local economies. A labor market with workers at all levels who are digitally proficient and able to engage in lifelong learning and job retraining is a community with widespread human capital that may also enjoy greater opportunity for its residents. It is also one that may be more resistant to the potentially harmful effect of artificial intelligence when jobs become automated.

Digital Human Capital

Broadband use can be conceptualized as a form of digital human capital. Economists often measure human capital in communities as educational attainment, especially the percentage of the population with college degrees (see Moretti 2012, 246). In what they call a digital human capital framework for communities, Bach and colleagues (2013) suggest that technology access and skills are analogous to formal education as a form of human capital.[6] Like universal public education, high-speed internet use in the population facilitates the development of skills and access to information that can enhance economic opportunity for individuals (Mossberger et al. 2003). Just as secondary schooling also influenced community prosperity in the industrial age (Goldin 1998), digital skills and information provide local capacity for innovation and opportunity today. Broadband use can be expected to add to the stock of human capital and to information flows in communities, enhancing local capacities for innovation and creating spillover benefits for the locality.

To follow we develop arguments for why broadband connectivity is critical in digital human capital for local communities. Broadband subscriptions can be viewed as promoting (1) access to information and capital-enhancing activities, (2) the development of skills, (3) multipliers and spillover effects, and (4) participation in information networks. The first two categories describe individual aspects of digital human capital, and the second two demonstrate how they may matter for the broader community.

Expanding Access to Information

Internet use may develop human capital through access to information. Hargittai (2002) has argued that some uses of the internet are capital-enhancing and likely to improve an individual's life chances. Internet use for education, job training, health, finances, and entrepreneurship are among the capital-enhancing activities that might be connected to economic opportunity or upward mobility (see also Hargittai and Hinnant 2008; Ragnedda 2018; Park 2017). Online education has been growing in the past decade for nondegree and degree programs (Koksal 2020). The internet offers educational choices without regard to location and is a critical resource for research and information in any educational setting. Job search is another obvious application where information can lead to economic advancement. Shearer and Shah (2018) identify promising industries where less educated workers can find entry-level jobs that are more likely to lead to opportunity in the future. This kind of career ladder requires the ability to search for and change jobs over time (Shearer and Shah 2018). Yet residents of low-income communities are also less likely to have information about good jobs in their personal networks, and the internet is a potential resource (Ioannides and Datcher Loury 2004). Thus, those who lack internet access and skills are excluded from the information needed to search for and apply for jobs that can enhance their human capital and outcomes.

Technology Use and Skills

Various skills have been identified in the literature on internet use, and these have evolved over time with technologies and devices. An early study by Mossberger and colleagues (2003) defined necessary skills as technical competencies, or the ability to use hardware and software, and information literacy, or the ability to search for, critically evaluate, and apply information. Others have identified operational skills for hardware and software, formal skills to navigate the internet, informational skills, communication skills, content creation skills, and strategic skills to pursue goals online or to improve one's position in society (van Deursen and van Dijk 2011, 2014; Schradie 2011). While broadband subscriptions do not indicate much about specific skills, personal internet use affords individuals more autonomy and the opportunity to develop skills (Hargittai and Hinnant 2008; Hassani 2006).

Such skills are in demand at the workplace. Laar et al. (2019) found in a survey of professionals that informational skills enabled use of the internet for problem-solving at work, with communications, content creation, and online collaboration as intermediate skills also facilitating problem-solving. While the aforementioned study focused on knowledge-intensive jobs, other occupations require digital skills as well, including jobs for less educated workers.

International data and research have focused on basic skills as core competencies needed in a variety of jobs, throughout the workforce. A report produced by Rand Europe for the UK government identified digital skills as both technical, on the one hand, and informational and strategic on the other (Grand-Clement 2017).[7] Compared with technical competencies, information skills and strategic knowledge about when to employ a technology solution are "eternal" and will survive changes in technology and enable self-learning and lifelong learning (Grand-Clement 2017). The ability to learn new skills over time is particularly important to keep up with rapid changes in technology and the evolution of work in the age of artificial intelligence (Karsten 2019).

Research utilizing OECD data from the Programme for the International Assessment of Adult Competencies (PIAAC) used a direct measure of internet skills in 19 countries and found that "ICT skills are highly valued in the labor market. . . . [I]f an average worker in the US increased their ICT skills to the level of an average worker in Japan, their wages would increase by about 8%; this is close to the well-identified estimates on the returns to one additional year of schooling in developed countries" (Falck 2017, 6). Similar to the Rand study, the author concluded that general competencies were more important than specific applications: "general ICT training may be a better way to prepare workers for future developments, providing them with a general toolbox that can be adjusted and honed to suit different situations" even as skills shift because of the Internet of Things and automation (Falck 2017, 9).

Community Impacts: Spillovers and Multipliers?

Longitudinal data shows that home internet use is associated with higher wages for individuals (DiMaggio and Bonikowski 2008). Controlling for other factors that influence wages, those who use the internet for their jobs earn more, including workers with a high school education or less (Mossberger, Tolbert, and McNeal 2008). There is reason to believe that this earlier research remains ever-more relevant in today's economy, including the results for less educated workers who can employ technology on the job. The Brookings Institution tracks "opportunity industries" that offer stable employment, middle-class wages, and benefits. While these industries may employ workers without a college degree, many jobs require some level of technology skill (Shearer and Shah 2018). Demand is projected to grow for these middle-skill jobs that require technology use but not college degrees, providing new opportunities for economic mobility (Horrigan 2018). Education remains a critical aspect of human capital in the knowledge economy, but internet use, like education, represents both skills that can be employed for economic benefit and access to information.

Does digital human capital—the ability to access and use IT throughout the community (Bach et al. 2013)—play a role in local economic outcomes? As with other forms of human capital, we expect that the skills and information imparted by broadband adoption in communities create regional spillovers and multiplier effects. In contrast with the share of IT jobs or other measures of the digital economy, broadband use in the population captures digital human capital in regions, cities, and neighborhoods; across sectors; for the highly skilled; and for less educated workers. Productivity gains from technology have been due to its application across occupations and industries (Brynjolfsson and Saunders 2010), and so the digital human capital of a region should contribute to productivity.

Some cross-sectional research using recent data from the ACS suggests that data on subscriptions is a predictor of local productivity (Gallardo et al. 2021). If more inclusive communities can draw upon a broader range of skills and deeper pools of talent (Florida 2017; Parilla 2017), then higher levels of broadband adoption should generate greater economic benefits.

Information Networks and Externalities

In addition to nurturing local human capital and its spillover benefits, broadband could be expected to develop information networks within communities and beyond their boundaries, encouraging the exchange of ideas for local innovation (Falck 2017). Major cities are innovators because of their dense networks of interaction and diversity of backgrounds and talents (Glaeser 2011, 6). Just as agglomeration economies convey advantages in physical space, virtual networks may foster interactions at the local level and connect communities across distances, introducing diverse experiences and knowledge. Integration into these larger networks may also have an impact on local outcomes (Falck 2017), supplementing face-to-face interactions in large cities, and filling resource gaps in smaller communities. In either case, positive externalities grow with participation in networks, and highly networked places, measured by their levels of broadband adoption, could be expected to generate more innovation and greater social benefit.

To summarize, broadband use in the population is a form of digital human capital that can enhance individual resources and promote spillover benefits for communities. Broadband subscriptions are a broader measure of technology capacity in a community beyond the location of Google and Amazon facilities or the development of gigabit networks. Widespread broadband use may offer other paths for fostering local prosperity in varied types of communities.

Prior Research on Broadband Outcomes

This study extends prior research on broadband's impacts on economic development in several important ways: it examines broadband subscriptions (use) rather than deployment (infrastructure), and population use rather than business applications by firms, testing the proposition that digital human capital creates local growth and economic opportunity. Overall, the literature has pointed to positive effects for broadband (Gillett et al. 2006; Kolko 2010; Jayakar and Park 2013; Atasoy 2013; Mack 2014; see also Falck 2017; Bertschek et al. 2016; and Abrardi and Cambini 2019 for international overviews), though there have been limitations in the data that have made it difficult to track broadband use. Data limitations have also made the establishment of causation problematic as well (Whitacre et al. 2014a; Gillett et al. 2006; Kolko 2010).

Most research on the impacts of broadband in the United States examines the effects of infrastructure deployment or service availability. Gillett et al. (2006) investigated deployment by zip codes across the nation and concluded that the introduction of broadband infrastructure was associated with faster growth in jobs and establishments, especially in IT-intensive firms. Crandall and colleagues (2007) noted positive relationships with employment across a range of economic sectors, including manufacturing, health care, finance, insurance, real estate, and education. Kolko's national study (2012) found an association between broadband deployment and economic growth, though this was most likely in industries that employ technology intensively and in places with less population density. Of particular note, however, was Kolko's conclusion that there was no relationship between deployment and local employment rates or wages, and therefore few benefits for residents. Kolko surmised that infrastructure investments attracted new workers with appropriate skills to the region. Atasoy's (2013) county-level research focused on labor effects between 1999 and 2007 and found that the impact of infrastructure was positive for employment of college-educated workers, negative for less skilled workers, and slightly greater for rural areas. Technology enables the displacement of jobs performing routine tasks (Atasoy 2013; see also Akerman et al. 2015 for evidence in Norway), but it is also possible that communities with more inclusive technology use can develop policies to promote skills needed for new occupations, such as the middle-skill jobs mentioned earlier. Falck (2017) points to international evidence that slow adoption of broadband by the population and limited technology skills in fact diminish the impact of infrastructure investments.

There are a few exceptions to the research on deployment, with studies using measures of broadband adoption or technology use in firms. Noteworthy research by Whitacre and colleagues (2014a) analyzed the role of broadband subscriptions (with the FCC quintile measure) side by side with infrastructure in nonmetropolitan counties. They found that over time, broadband adoption was a stronger predictor of economic outcomes than deployment. Forman and colleagues (2012) concluded that internet use by firms leads to positive outcomes, but only in 6% of metropolitan counties. These were counties with the largest populations, greatest wealth, and highest technology industry presence. We will review these studies further in the chapters on counties and metropolitan regions.

While the economic impacts of broadband infrastructure overall have been positive, previous research shows they have varied across places in the benefits for residents, including wages and employment. *No previous research, however, has analyzed the effect of broadband subscriptions on economic outcomes for urban and rural communities over time.*

Overview: Data, Methods, and Trends in Broadband Subscriptions, 2000–2017

The US Census Bureau CPS has tracked internet use since 1997, and broadband or high-speed adoption since 2000.[8] National data from the CPS and state-level estimates have been reported by the US Department of Commerce (NTIA Data Central 2020), but no government data was available below the state level until the ACS began to ask three questions on computer and internet use in 2013. The question on broadband adoption asks respondents: "At this house, apartment or mobile home, do you or any member of this household subscribe to the internet using (a) dial-up, (b) DSL, (c) cable modem service, (d) fiber-optic service, (e) mobile broadband plan for a computer or cell phone, (f) satellite service, (g) some other service?" Responses (b)–(f), including mobile and satellite, are counted in federal data as broadband (dial-up is not broadband) and in our measure of broadband subscriptions. There are differences between these types of access, however, and we discuss smartphone-only internet use in Chapter 4. The questions asked in the CPS from 2000 to 2012 are comparable to the ACS, though they have varied slightly over time. A table with the questions and descriptions of how they were coded across years is available in Appendix A.

Methods and Data for Broadband over Time

We provide a comprehensive view of broadband subscriptions or use during a period of rapid growth, when the internet evolved from an emergent to a dominant communications technology in society. We develop data on broadband subscriptions for counties and metros from 2000 to 2017, drawing on the 2000–2012 CPS of approximately 100,000 respondents per year, creating estimates with multilevel models and poststratification weights. We also use available data for counties, metros, and states from the 3-million-person annual ACS from 2013 to 2017 to create local community estimates. Combined, the data utilizes individual survey responses from millions of Americans over nearly two decades to create a historical timeline of the geography of broadband diffusion across the United States.

To generalize from the small samples for many of these metros and counties in the 2000–2012 CPS, simple disaggregation could have led to bias. To overcome this problem, we use multilevel or hierarchical linear modeling to estimate broadband use for states, counties, and metros during this period (Mossberger et al. 2013). Multilevel regression with imputation and poststratification weights to create population estimates for geographic areas is more precise and robust than survey disaggregation, leading to less error and narrower confidence intervals in our predictions (Lax and Phillips 2009a, 2009b; Park et al. 2004, 2006; Pacheco 2011). Individual-level variables measuring demographic factors (race/ethnicity, gender, education, income, age, 16 occupational categories from the North American Industry Classification System, etc.) are merged with aggregate-level variables measuring the percent unemployed or the percent in poverty for metros and counties. We also include population size for the county estimates. With this multilevel data (individuals nested in local communities) and the statistical method using probability simulations, we model broadband subscriptions for local geographic areas. Community-level estimates for each year are generated separately and then combined to create the time series.

In the remainder of this section, we explore this data to identify general trends that have emerged in broadband adoption over time, across communities.

Broadband Adoption Has Plateaued across Geographies

One feature apparent in the longitudinal data is a leveling-off of broadband adoption since 2010–2011, despite federal efforts to increase broadband deployment and adoption during the stimulus programs following the 2008 recession. Broadband adoption rates have flattened across geographies, with nearly 20 million households (17%) lacking broadband subscriptions of any type, including smartphones.

Just as national data has shown a leveling-off for broadband subscriptions (Pew Research 2019; NTIA Data Central 2020), the same is true across geographic scales, even though some rural counties have had further to go to catch up. Figures 1.3, 1.4, and 1.5 show similar trends across states, metros, and counties.

Localized Data Reveals Wide and Persistent Inequalities

There are larger gaps in broadband subscriptions across local communities in the United States than apparent in national survey data, which previously identified differences for urban, suburban, and rural respondents at best. Not all urban or rural communities have similar broadband adoption rates. And aggregation to the state level is inadequate to portray differences

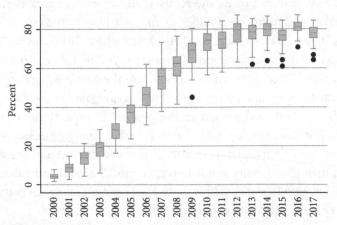

Figure 1.3 Percentage of the population with a broadband subscription, 2000–2017, for states. Data from 2000–2012 is derived from CPS; 2013–2017 is drawn from the ACS.

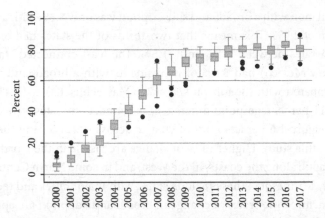

Figure 1.4 Percentage of the population with a broadband subscription, 2000–2017, for counties.

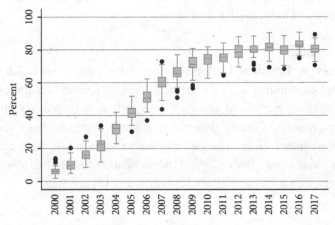

Figure 1.5 Percentage of the population with a broadband subscription, 2000–2017, for US metros.

across local communities within the state. The longitudinal data shown here demonstrates that gaps across places have been persistent, and they continue currently, even taking mobile access into account.

State Trends

Disparities in subscription rates across the states are greater than across metro areas because they include rural communities. Still, they are smaller than gaps evident across counties. In 2001, the average state percentage of the population with a broadband subscription was only 8.8% with a standard deviation of 2.7%. This means that two-thirds of the states had broadband adoption rates between 6.1% and 11.5%. The least connected state at the time (New Mexico) had 3.5% of its residents with a broadband subscription, compared with the top-ranked state (Massachusetts) at 14.4%. This is only an 11-percentage-point difference, as connectivity rates were quite low overall. Figure 1.6 depicts rates of change across states during the 17-year period in this study. Higher rates of change are visible in the Northeast and Mid-Atlantic along the coasts, in the West, and in some North Central states, especially Minnesota. These include states with major cities and technology employment. Slower rates of change in broadband adoption are apparent in areas of the South and Southwest. Michigan's growth stands out as particularly low for the Upper Midwest. Conversely, Georgia exceeds the rate of change in its region, possibly due to Atlanta's growth. Yet Texas, with tech hub Austin and other major cities, is relatively low in broadband growth, perhaps because of its sizable rural expanse.

County Trends

Disparities in broadband adoption from the most connected to least connected communities are the greatest across the nation's counties, which include sparsely populated rural areas as well as high-poverty and high-minority counties. County differences are much greater than across the 50 states or metro areas. The gaps between the most and least connected counties range from a low of 25% of the population to a high of 95+%. There are striking disparities in digital adoption across US counties.

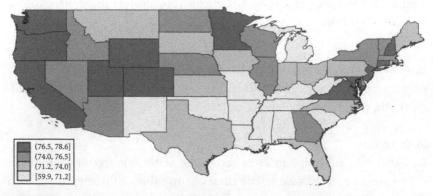

(76.5, 78.6]
(74.0, 76.5]
(71.2, 74.0]
[59.9, 71.2]

Figure 1.6 State change in broadband subscriptions, 2000–2017.

In 2001, average county home broadband adoption was only 10.4%, with a standard deviation of 4.3%. Most counties had broadband subscription rates between 6.1% and 14.7%. The least connected county (Bay County, Florida) had 6.9% of its population with broadband, and the most connected county (Howard County, Maryland) had 46.8% of its population with broadband subscriptions. By 2017, 95% of the population in Douglas County, Colorado, had broadband subscriptions, but this compared to only 24% in lowest-ranked Wheeler County, Georgia.

There are also dramatic differences across US counties in the percentage of the population with high-speed internet depending on family income (2017 ACS). The three maps in Figures 1.7, 1.8, and 1.9 show the percentage of the population with a broadband subscription (including mobile only) for low-income households earning less than $20,000 per year, middle-income households earning $20,000 to $75,000 per year, and high-income households earning more than $75,000 annually. In the West, even low- and middle-income households are more connected than in other parts of the country. Affluent individuals have high levels of internet connectivity nationwide, across regions and counties. These maps are revealing, for they expose not only the substantial regional variation but also that inequality in information access tracks closely with economic inequality in general.

Metropolitan Trends

Differences in broadband subscriptions for the least and most connected communities are the smallest across metropolitan areas, as suburban

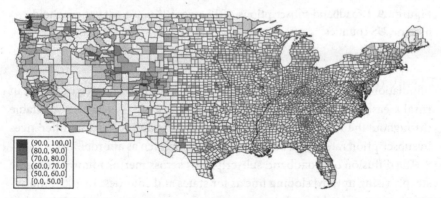

(90.0, 100.0]
(80.0, 90.0]
(70.0, 80.0]
(60.0, 70.0]
(50.0, 60.0]
[0.0, 50.0]

Figure 1.7 Broadband subscriptions, 2017, less than $20,000 median household income, US counties.

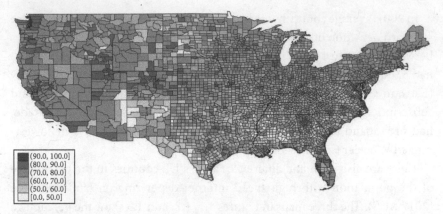

Figure 1.8 Broadband subscriptions, 2017, $20,000–$75,000 median household income, US counties.

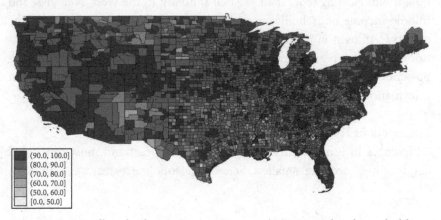

Figure 1.9 Broadband subscriptions, 2017, over $75,000 median household income, US counties.

populations included in metro data tend to have higher rates of use than rural areas or central cities. Broadband infrastructure is generally available throughout the 50 largest metropolitan areas we examine here, so differences in subscription rates reflect barriers to adoption such as affordability.

The diffusion of broadband subscriptions across metros follows the same steeply rising upward sloping line as for states and counties. In 2001, the average percentage of broadband adoption for metros was just 10.8%, with a standard deviation of 3%, meaning that around 68% of metros had between

Figure 1.10 Change in broadband subscriptions, 2000–2017, US metros.

7.8% and 13.8% of their residents with broadband subscriptions. The least connected metro (Miami–Fort Lauderdale–West Palm Beach, Florida) had 5.7% broadband adoption, and the most connected metro (San Diego–Carlsbad, California) had 20.3%. The total range was about 14 percentage points. In 2017, the San Jose–Sunnyvale–Santa Clara, California, metro had the highest rate of broadband adoption at 89%, whereas the Brownsville-Harlingen, Texas, metro had the lowest rate of broadband subscriptions at 53%. Metropolitan differences today are much wider, with a 36-percentage-point difference. Figure 1.10 demonstrates the variation in the growth of broadband subscription rates, with some cities seeing much larger changes than others from 2000 to 2017.

Central Cities
Many central cities and low-income suburbs have substantially lower rates of broadband adoption than their surrounding metros. For example, in the 50 largest cities that we study here, broadband subscriptions including cell phones ranged from 92% in Raleigh, North Carolina, to 66% in Miami in 2017. For fixed broadband not including cell phones, Seattle had the highest percentage of subscriptions at 85%, compared to 48% in Detroit. Chapter 4 will discuss variation in broadband adoption in central cities, including differences across neighborhoods and demographic groups within the central cities of the 50 largest metros.

   In Chapter 4 we explore new data on neighborhoods within cities, portraying the disparities that exist even within highly networked places. There is insufficient data to include neighborhoods in the time series here, but we use the 2017 ACS to illustrate the inequalities that exist across neighborhoods (Figure 1.11). A study conducted by the Brookings Institution (Tomer et al. 2017) used the tract-level data from the FCC to show how the quintiles for broadband subscriptions varied by community income. Clearly broadband subscriptions are related to neighborhood income as well as residential segregation, and low rates of adoption may further limit opportunities for economic mobility.

Given the role that technology has played in the economy, examining trends and disparities in broadband subscriptions is critical for understanding the past and the present of local communities in the United States. Beyond these trends, we examine the role that broadband adoption has played in community prosperity and growth. Better evidence is needed for policy, and for community leadership to address both innovation and inequality in the rapidly unfolding digital economy.

Organization and Plan of the Book

The ensuing chapters explore how broadband connectivity in the population affects economic opportunity across places, using nearly two decades of new community-level data.

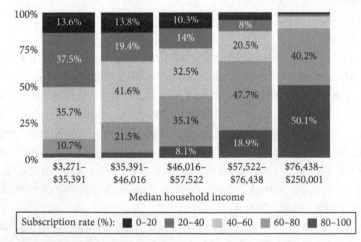

Figure 1.11 Broadband subscriptions by neighborhood, 2016.
Source: Tomer et al. (2017), Brookings Institution.

Chapter 2 provides a lens to examine urban and rural divides in broadband use, and in community well-being and opportunity, drawing on the nation's 3,100+ counties. Rural areas are more likely to lack adequate broadband infrastructure and are also among the communities most likely to experience economic distress. We invert the Economic Innovation Group's Distressed Communities Index as a measure of community prosperity in counties. Changes in broadband subscriptions and absolute broadband rates (with lagged time series models) are used to predict economic outcomes (growth in median income) in both urban and rural counties over time.

Metropolitan areas represent regional labor markets and the drivers of the national economy, and it is critical to examine outcomes at this level. While some earlier studies have studied the effects of internet use by firms in metropolitan areas (Forman et al. 2012), no previous research has definitively modeled the impact of broadband use by the population for metro areas over time. Chapter 3 explores the extent to which broadband subscriptions contribute to metropolitan growth and prosperity, using annual measures developed by the Brookings Metro Monitor (Shearer et al. 2018). Time series analysis and lagged terms are again used to model how change in broadband rates and absolute broadband adoption are related to growth and prosperity in the nation's 50 largest metros.

Chapter 4 analyzes patterns of broadband use in the central cities at the core of metropolitan areas, ranking them and identifying demographic differences in broadband adoption. This chapter exploits 2017 broadband data that allows us to track neighborhood variation within cities (at the zip code level). Just as Chetty and colleagues have argued that opportunity is influenced by the local context, down to the level of the neighborhood, so it is necessary to explore broadband adoption at this scale. Local governments and nonprofits are often engaged in initiatives to promote greater inclusion in targeted areas, and we close the chapter with an evaluation of neighborhood-level change in Chicago's Smart Communities.

States shape the policy context for communities in many ways. If we think of broadband use in the population as an information resource, what is the role of technology in state policy innovation over time? Do more inclusive information networks lead to increased innovation? Is this a spillover effect of greater digital human capital in the state population? By including broadband subscriptions in a model of diffusion in Chapter 5, we examine how the internet's dramatic expansion of information flows affects policy in the states. The State Policy Innovation Dataset (SPID) of thousands of policy adoptions from 2000 to 2017 and our state measures of broadband adoption are used in a pooled event history model (Kreitzer and Boehmke 2016) to estimate the relationship between broadband adoption and policy innovation. Our analysis uses changes in broadband adoption and absolute broadband adoption rates (with lagged time series models) to predict state policy innovation over time and the environment for local innovation.

In Chapter 6, the final chapter, we return to the two place-based narratives of innovation and inequality and consider the possibilities for reconciling the impacts of technology for communities. As we build "smart" cities and places, employing artificial intelligence and the Internet of Things; as economies and jobs continue to experience rapid change; and as more information and services migrate online, our analysis shows it is communities with widespread broadband use that will best exploit these new technologies and other innovations. Such communities have already proven more resilient in the face of COVID-19. Counties with a higher rate of broadband subscriptions combined with more commercial domain name websites per 100 had lower unemployment in April 2020. This was the case even for counties in the top 50% for COVID-19 cases. In closing, we present a research agenda and policy choices for a future that enables truly smart and inclusive innovation, where communities have the digital capacity to thrive.

2

Counties

Broadband Use and Prosperity across Diverse Contexts

> Access to affordable, high-speed internet is essential to connect
> people and places and compete in today's economy.
>   —Matthew Chase, Executive Director of
>   the National Association of Counties (NACo) (LISC 2019)

Introduction

Rural Yancey and Mitchell Counties in North Carolina formed a public-
private partnership with Country Cablevision to install a fiber-optic network
that offers some of the fastest speeds in the state—up to 100 Mbps for residential service and 1 gigabyte for businesses. Prior to this, according to the
School of Government at the University of North Carolina:

> Local businesses struggled, and a local Glen Raven manufacturing plant
> was on the brink of closing because of the lack of broadband. Attracting
> and retaining employees with appropriate skills was becoming difficult for
> local manufacturers like BRP and educational institutions in the commu-
> nity were not able to provide programs that could train local students for
> these jobs. Tourism also suffered as hotels, restaurants, and more were un-
> able to attract customers, run their business, or provide internet services for
> their guests. (Efird 2020)

A $25 million grant from the US Department of Agriculture in 2010 launched
the bi-county broadband initiative, which has since expanded the network
to cover 97% of homes and businesses. Federal funds and private partners
were needed for the project, and implementation was complex because of
pre-emptive legislation in North Carolina that prohibits local governments
from paying for or regulating broadband networks. Yet once in place the

Choosing the Future. Karen Mossberger, Caroline J. Tolbert, and Scott J. LaCombe, Oxford University Press. © Oxford
University Press 2021. DOI: 10.1093/oso/9780197585757.003.0002

broadband networks facilitated collaboration between the community college and a local manufacturer for an advanced manufacturing training program. Small businesses such as restaurants reported that the networks allow them to use business logistics, point-of-sale networks, and free Wi-Fi for customers (Efird 2020).

Infrastructure was a central issue in Yancey and Mitchell Counties, but investments are often related to business calculations about the ability of local populations to pay as well as the expense of deployment in sparsely populated or geographically challenging areas. Nationally, only 13% of rural counties with median incomes of $75,000 or more lack broadband access (Efird 2020). While some rural county governments have struggled to attract broadband service providers, urban counties have also faced disparities in broadband adoption and use.

In Cuyahoga County, which includes the city of Cleveland, approximately one-quarter of the county's population had no internet subscription at all in 2017, either cell or wireline broadband, and that rises to one-half of the population with annual household incomes of $20,000 or less (Schartman-Cycyk et al. 2019, 1). Social programs administered by the county serve many of the low-income residents and seniors who are least likely to be online, and greater broadband adoption could promote more efficient service delivery and economic opportunities for residents. The county is collaborating with the nonprofit DigitalC to pilot affordable fixed wireless broadband in one community, with a view toward expansion of the program in other underserved city and suburban neighborhoods. "Internet access is critical to day-to-day life," said County Executive Armond Budish in the county's official statement announcing this collaboration. "We are aiming to provide in-home broadband access to our residents so that they can more easily access services . . . and make our community more prosperous" (Cuyahoga County 2019).

Counties provide a critical lens for measuring technology use and economic results in rural and urban areas simultaneously.[1] Additionally, throughout the United States, counties are the primary providers of direct social services to families, children, and adults. Counties promote community well-being through food subsidies, health care, job training, subsidized housing, adoption services, and community management and infrastructure. Counties are also generally engaged in local economic development efforts, and so they are responsible for many activities affecting regional prosperity for residents and businesses (NACo 2019).

This chapter focuses on US counties and provides evidence that broadband adoption affects local prosperity, using instrumental variable models and time series analysis as well as cross-sectional models. With new data on all counties as well as longitudinal data, we are able to explore for the first time outcomes for broadband adoption in all types of counties rather than the deployment of infrastructure. We explore the impacts of digital human capital for inclusive growth that benefits both residents and businesses across urban, suburban, and rural counties.

Surveying the County Landscape

Nationwide there are 3,142 counties or county-equivalent governments (Louisiana uses parishes instead of counties and Alaska has boroughs). Variation across counties is vast; Texas has 254 counties, more than any other state, while Delaware has 3, the fewest of any state. Washington DC County is home to the District of Columbia. The number of counties per state is not constant. Iowa, with a population of just over 3 million people, has 100 counties, while California, home to nearly 40 million, has 58 counties. Not only does the number of counties per state vary, but also so do their populations, ranging from Los Angeles County with over 9.9 million people and Cook County, Illinois (home to Chicago), with 5.2 million to Kalawao County, Hawaii, with 89 residents in 2015. Nearly 1,000 US counties have a population over 50,000 (American Community Survey [ACS] five-year estimates).

Recent census data indicates that 31% of the population resides in urban counties, with 55% and 14% in suburban and rural counties, respectively (Figure 2.1). There are 68 urban counties—for example, Miami-Dade County, Milwaukee County, and San Diego County as well as Cuyahoga County— located in the 53 US metropolitan areas with at least a million people. There are 1,093 suburban and small metro counties—outside the core cities of the largest metro areas, as well as the entirety of small metro areas. Home to just 14% of the US population, the majority of counties (1,969) are rural, located in nonmetro areas and with a median population size of 16,535 (Parker et al. 2018). Since 2000 a shrinking share of the population lives in rural areas (3% population change), while suburban area population is growing (16%) and urban areas are holding constant (13%).

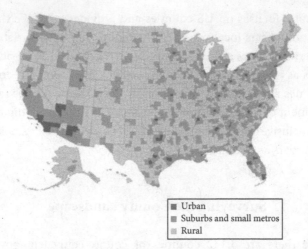

Figure 2.1 Geographic distribution of urban, rural, and suburban counties, 2017 American Community Survey (ACS) one-year estimates, including broadband and satellite.

Surveying the county landscape captures the diversity of communities in the United States, with wide variation in the local context as well as differing population densities. Recent data from the US Census Bureau's ACS and a unique time series dataset developed by the authors demonstrates that broadband subscriptions vary widely across counties, with substantial gaps in the connectivity of the population between cities and regions, and especially between urban and rural places. Rural counties still lag considerably behind suburban and urban counties in broadband adoption (Mossberger et al. 2013). In 2017, Douglas County, Colorado, had the highest rate of household broadband subscriptions in the nation at 96%. A suburban county within the Denver metropolitan area, Douglas had the lowest jobless rate in the region and was ranked by the Center for Digital Government as one of the top 10 US county governments for digital innovation.[2] In contrast, Apache County in Arizona had just under 40% of the population with some type of broadband subscription in 2017. The rural county includes both Navajo and Apache tribal communities, with two-thirds of the county's population and over one-half of its land area belonging to the Navajo Nation.[3] What

is the impact of these spatial inequalities for resident and business opportunities in these communities?

Comparing Prosperity and Distress in US Counties

Economic well-being varies across communities in the United States,
and we explore how this is related to broadband adoption. While prosperous counties have become richer, distressed places have become poorer
and the gaps between the two have widened, according to the Distressed
Communities Index (DCI) compiled by a think tank called the Economic
Innovation Group (2018). In this chapter, we measure economic prosperity
(a modification of the DCI), change in economic prosperity over time, and
growth of median household income for US counties.

The DCI places communities (counties and zip codes) on a single index
to compare measures of prosperity and distress (Economic Innovation
Group 2018). It is composed of seven components: the percentage of adults
without a high school diploma, housing vacancy rate, percentage of adults
(ages 25–64) not in the workforce, poverty rate, median household income
adjusted for state income, and change in the number of jobs and number
of businesses (2012–2016). The authors consider the top one-fifth of communities "prosperous," defining the second quintile as "comfortable," the
third "mid-tier," the fourth "at risk," and the fifth, or worst off, as "distressed."
Following the 2008 recession, counties in the top quintile have thrived with
significant growth in jobs, businesses, and population, while the number of
people living in economically distressed counties (the bottom quintile) have
shrunk and more places in the distressed category are rural. This reflects
other recent trends. Nationwide a third of Americans say they don't have
enough income and do not expect to in the future, but this rises to 40% of
people living in rural counties and 36% in urban counties (Parker et al. 2018).

In an increasingly digital economy, rural counties were more likely to be
downwardly mobile following the 2008 recession. The Economic Innovation
Group includes comparisons between two time periods: 2007–2011, just before and during the height of the last recession, and 2012–2016, during the
years of recovery. The average state had 27% of its population living in a prosperous zip code from 2012 to 2016, compared to 14.5% living in a distressed

zip code. Regional patterns are strong, as Southern states had much higher percentages of their population living in distressed places. In Alabama, Arkansas, Mississippi, and West Virginia, one-third or more of the population resided in a distressed zip code. Growth in new businesses and jobs was unevenly distributed across counties. Since the 2008 recession, highly populous counties—with more than half a million residents—were far more likely to add businesses when comparing the prerecession to postrecession period. Some places, such as the rising innovation hubs of Salt Lake/Provo, Utah, and Denver, Colorado, performed particularly well after the recession; Utah led the nation in community economic prosperity, with half its population residing in a top-tier zip code. Colorado followed closely behind. After the recession, Utah's share of the population living in prosperous zip codes increased by 18 percentage points, with similar advances throughout much of the West. In California, 3 million more residents lived in prosperous zip codes after the recession than before (Economic Innovation Group 2018).

How is broadband adoption and skill in the general population associated with trends in prosperity and distress over the past two decades? Controlling for other known predictors of economic development, what role does technology use in the population play in community outcomes? *This chapter examines how broadband adoption matters for policy-relevant outcomes, such as economic growth and prosperity, over the past two decades. Examining trends over time is essential for developing causal arguments in social science, where change in the predictor variable precedes the outcomes measured.* Tracking broadband adoption over time in local communities in the United States sheds light on the needs and resources to address both innovation and inequality in the rapidly changing digital economy.

Rarely in studies of local economic development has digital human capital been a primary factor in the analysis. While some prior research has shown that broadband infrastructure affects economic outcomes in counties, evidence on broadband use (i.e., subscriptions) has been limited because of the lack of adequate data. One interesting study of rural counties, however, found that broadband adoption (the percentage of the population with broadband at home) is a better predictor of economic outcomes than broadband infrastructure/availability of service (Whitacre et al. 2014a). This chapter hypothesizes that digital skills and information contribute to human capital in both urban and rural communities; this may be more important than previously recognized for combatting spatial inequality in the digital economy and creating opportunity.

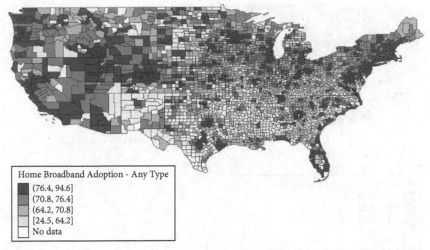

Figure 2.2 Percent of population with a home broadband subscription (mobile or wired), by county, 2017 American Community Survey (ACS) five-year estimates.

Measuring Broadband Subscriptions across Counties

For the first time, we have a nationwide view of broadband subscriptions for all 3,100 counties, with new tract-level data released by the US Census Bureau in December 2018 for the 2017 ACS. Figure 2.2 shows the percentage of households in a county with a broadband subscription of any type, while Figure 2.3 omits the nearly 20% of broadband users who rely on mobile-only internet. The pattern of connectivity is similar across counties, ranging from places where just one in four households have high-speed internet to counties where over 90% of the households are online. The maps show remarkable differences and are long overdue, given that broadband has diffused for nearly two decades now. Prior to 2018, data to measure broadband use was not available for 75% of US counties.[4]

Figure 2.2 shows that broadband subscriptions (including both wireline and cell) are highest in counties near the East and West Coasts, and in some areas of the Mountain West, the lower Great Lakes, and metropolitan areas in Texas and Minnesota. Counties with higher rates of broadband adoption are shaded darker. Notably, counties in the South and parts of the Midwest are more digitally disadvantaged, as are some places in the rural West. Much of New Mexico and the eastern part of Arizona are visibly lower, including the

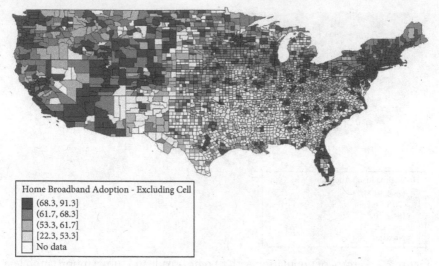

Figure 2.3 Percent of population with a home broadband subscription (excluding mobile only), by county, 2017 American Community Survey (ACS) five-year estimates.

Navajo Nation and some other tribal lands. Central Florida and some areas along the Canadian border have sparser adoption than surrounding regions. While most counties in southeast Michigan rank in the top quartile, Detroit's Wayne County does not. Urbanized areas in the South, such as Atlanta's metropolitan counties, stand out as more connected than much of the region.

Figure 2.3, which excludes the population that relies exclusively on cell phones for connectivity, is largely similar, though rates of connectivity range from 22.3% to 91.3%. This contrasts with Figure 2.2, where broadband subscriptions (including cell and wireline) range from 24.5% to 94.6% of the population.

Table 2.1 shows the timeline for when data to measure broadband subscriptions for counties became available from the government. The Computer and Internet Use supplements for the annual Current Population Survey (CPS) of the US Census Bureau first included questions in some years on internet subscriptions in 1997 and home broadband in 2000 (cable, DSL, cellular, satellite, fiber, etc.) (NTIA 1998, 2000). These data, with a 100,000-person sample, were used to develop reports of the US adult population online for the nation as a whole, but no data was available to describe the substantial local and regional variation, within and across communities.

Table 2.1 Timeline of Data Availability to Measure Broadband Subscriptions for Counties

American Community Survey (ACS) One-Year Estimates (small population counties where data is not available shaded in gray)	Broadband Subscription Data	Number of Counties with Data	Number of Counties without Data
	2000–2012 Author's estimates from US Census Bureau Current Population Survey	324 largest-population counties	2,776
	2013–present US Census Bureau ACS	800 counties with population over 65,000	2,300
	2017 US Census Bureau five-year ACS	3,100	0

Given the lack of broadband data for local communities, a prior National Science Foundation grant developed by the authors used the CPS data and hierarchical statistical modeling to create the first available measures of broadband subscriptions for the largest-population counties (324) from 2000 to 2017 to fill in the gaps (see Broadband Data Portal).

It was not until 2013 that the ACS, with its 3-million-person annual sample, first included questions modeled after the earlier CPS to measure broadband adoption, making data on broadband use available for the nation's 800 largest-population counties. Geographic identifiers for individuals living in places with populations smaller than 65,000 people were missing from the one-year census estimates, so there was still no data for most counties even after 2013 (see Table 2.1). Still, for the 800 counties that could be estimated from the ACS one-year data, there were geographically large but sparsely populated counties, especially in the West.

In December 2018, the US Census Bureau released the 2013–2017 five-year estimates from the ACS providing the first comprehensive dataset

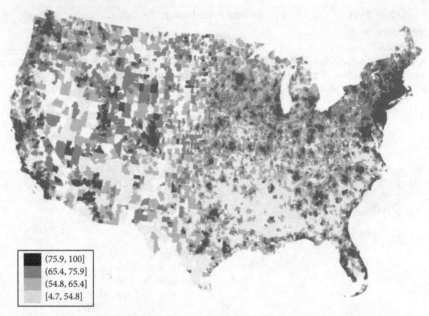

Figure 2.4 Broadband subscriptions (percent excluding mobile only) per zip code, 2017 American Community Survey (ACS) five-year estimates, census.

of the percentage of the population with home broadband subscriptions for the nation's 3,100 counties, 30,000 populated zip codes (Figure 2.4), and 73,000 census tracts (neighborhoods). While complete, the data is for just one year. Before this time connectivity in most counties went unmeasured, along with the nation's zip codes and census tracts. This lack of basic data on population broadband use is a concern for research and policy.

This study draws on newly developed time series data for counties from 2007 to 2017 from a subsample (the larger population counties), as well as cross-sectional data for 2017 only using the complete set of data from approximately 3,100 counties. The map in Figure 2.4 aggregates the five-year estimates to the zip code level for broadband subscriptions (excluding cell phones). Compared to the county maps, this shows somewhat more variation. This is especially true in the West, where the large geographic size of the counties distorts the extent of connectivity. The zip code map shows to a greater extent the concentration in more urbanized or populated areas. Still, many areas in the West as well as along the coasts have higher rates of broadband adoption.

Prior Research on Broadband Impacts

Why would broadband matter for economic outcomes? Prior research demonstrates that there are community-level benefits for the availability of broadband infrastructure or for technology use by businesses. Research on broadband infrastructure has associated it with community benefits such as faster job and firm growth (Gillett et al. 2006) and employment across a range of economic sectors, including manufacturing, health care, finance, insurance, real estate, and education (Crandall et al. 2007). Causal relationships have been difficult to establish, however (Whitacre et al. 2014a).

Some evidence has been mixed in terms of the impact that broadband infrastructure or availability has for local residents. Kolko's (2012) national study of broadband deployment using zip code–level data revealed outcomes such as economic growth, but without benefits such as increased wages or employment for local populations. In other studies, broadband infrastructure has been associated with decreased county unemployment rates (Jayakar and Park 2013) and other economic benefits (see Holt and Jamison 2009 for a review). One study of broadband availability in US counties from 1999 to 2007 found that it had a significant and causal effect on employment mostly through expansion in existing firms (Atasoy 2013). The author found, however, that the effects were positive only for workers with college degrees, and significant and negative for noncollege workers. The impact was greater in counties with more skilled workers and in industries requiring more education, and rural areas benefited somewhat more, though the differences were modest (Atasoy 2013). Similarly, Mack and Faggian (2013) examined counties between 1999 and 2007 and found positive effects for labor force productivity of broadband infrastructure overall, but negative impacts for noncollege workers. They recommend attention to digital skills and retraining for workers with routine tasks being replaced by technology.

Some studies have addressed aspects of digital human capital across different geographies. With firm-level data, Forman et al. (2012) found that advanced business uses of the internet enhanced productivity and were related to wage and employment growth, but only in large counties with a high presence of information technology (IT) industry. Clearly agglomeration effects mattered, and the authors were able to examine the different ways that firms used broadband. None of these studies, however, addressed the impact of broadband adoption by the population, which is a better measure of digital human capital throughout the community.

In contrast with prior studies, Whitacre and colleagues (2014a) used propensity score matching to study the role of broadband adoption as well as infrastructure and speed in rural counties, measuring change between 2001 and 2010. Their study found that high levels of broadband adoption predicted growth in median income and the number of firms as well as reductions in unemployment, whereas measures of broadband availability had limited effects on outcomes.[5] Counties with low levels of broadband adoption experienced declines in the number of firms and rising unemployment over this period. In addition to focusing on nonmetropolitan counties only, their study relied on the categorical data that the Federal Communications Commission (FCC) made available at the census tract level beginning in 2008. Following the release of the more precise ACS data, Gallardo and colleagues (2021) examined various measures of broadband availability and use, assessing their impacts on productivity for all counties. They found that broadband subscriptions and ACS data on devices (an additional measure of digital capacity) were better predictors of labor productivity than availability. While limited to relationships measured in a single year, the effects for productivity suggest digital human capital as a factor, though there were some differences for urban and rural counties.

How is it that broadband connectivity and use might affect community prosperity in rural counties, in particular? Broadband use has the potential to break down geographic barriers for communities that are isolated from the hum of economic activity in metropolitan areas. One way is by connecting local businesses to suppliers and markets through websites, social media, email, cloud computing, data, online research, and more. Broadband may be used for information sharing, purchase channels, or sales channels (Stenberg et al. 2009). This can connect brick-and-mortar businesses or even entrepreneurs selling crafts from their homes to broader markets. Workers with home broadband can search for jobs within commuting distance or even in other cities, supplementing their own personal or localized information networks. Reliable broadband connections also open possibilities for telework. This has been growing in recent years but is expected to increase even after the COVID-19 pandemic, with companies announcing that remote work will be a longer-term option (Gaskell 2020). Broadband also facilitates distance education, from vocational training to college degrees (Stenberg et al. 2009). Thus, broadband subscriptions may contribute to digital human capital through information access for entrepreneurship and employment.

It may promote the skills of the rural workforce. These developments should have positive externalities in the community and support richer information networks for local innovation.

More evidence is needed for urban and rural communities, at different scales. Broadband adoption, in contrast with availability, may be low in some urban or suburban communities because affordability is a barrier to adoption. Concentrated poverty, which had declined in the early 2000s, has grown since the 2008 recession. It is spreading in suburban communities and in regions of the country outside its traditional strongholds in the large cities of the Northeast and Midwest (Kneebone and Holmes 2016; Allard 2017, 5).

Growing economic inequality across communities raises questions about potential paths forward in the digital economy. One question is whether broadband adoption in a county benefits mainly college-educated workers or the broader community. The map in Figure 2.5 shows counties ranked by the Distressed Communities Index with data from the Economic Innovation Group for the nation's 3,100 counties (also available for 30,000 zip codes). We have inverted the scores to construct an Economic Prosperity Index so that higher values (marked in darker shades) represent prosperous areas. Scores range from 0 to 100, with 100 representing the highest level of prosperity. The annual index is based on seven core components including poverty rates,

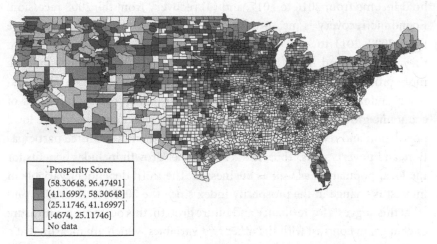

Prosperity Score
(58.30648, 96.47491]
(41.16997, 58.30648]
(25.11746, 41.16997]
[.4674, 25.11746]
No data

Figure 2.5 Economic prosperity by county (darker = more prosperous), Economic Innovation Group.

workforce participation, educational attainment, housing vacancy, change in jobs, median income, and change in business establishments.

Rural areas and regions such as the South tend to be the most economically distressed. The most prosperous counties saw a 20% increase in employment from 2012 to 2016, compared to the most distressed. And critically important for human capital, the mean percentage of college-educated residents is just 17% in the most economically distressed places; it more than doubles (to 36%) in economically robust counties when ranking counties by quartiles on the index. There is a large spatial component for social and economic disadvantage in the United States.

Data Analysis: Predicting County Economic Prosperity with Broadband Subscriptions

Can broadband connectivity and use improve a community's economic well-being? We examine this first using the 2017 ACS five-year estimates on broadband subscriptions for the nation's 3,100 counties as the primary predictor variable in a multivariate regression model. This provides a view of all counties that was impossible before. We use the 2017 five-year estimates to predict three county-level economic outcomes: (1) scores on the economic prosperity index (shown in Figure 2.5), (2) change in median household income from 2016 to 2017, and (3) recovery from the 2008 recession. Economic recovery is measured as change in the county prosperity score from 2007–2011 to 2012–2016. We also update the outcome variables so they are measured in 2018 with similar findings. Positive values indicate more prosperity and growth (recovery) following the recession. The prosperity index, described earlier in this chapter, encompasses indicators of economic growth, such as change in businesses, and outcomes for the local population and region, such as full-time employment, workforce participation, and poverty rates. Thus, it tracks whether growth includes benefits for the local population as well as businesses. The third dependent variable of interest is change in the prosperity index since the 2008 recession. Beyond what this suggests for resilience and future growth, this outcome is a measure of change, in contrast with the other two variables, which are measured at a single point in time.

To control for other factors related to economic outcomes, the statistical models include variables to measure the educational attainment of communities with covariates for percentage of high school and college graduates (less than high school is the reference group). As mentioned in Chapter 1, educational attainment is a commonly used measure of human capital and is clearly related to economic outcomes in prior research on broadband (Atasoy 2013; Mack and Faggian 2013). The local workforce is shaped by the age of the population, and models include age cohorts[6] (silent generation, those born before 1946, is the reference category). Less prosperous communities have a higher share of minorities (Economic Innovation Group 2018), and models account for racial and ethnic group population size (white non-Hispanics is the reference group). Data from the US census North American Industry Classification System (NAICS) codes are used to control for the percentage of the population employed in different industries (with manufacturing as the reference group). The latter includes the percentage of the population employed in IT jobs. The regression models are weighted by population given the diversity among counties.

Because we are interested in the impact of digital human capital, we introduce another measure of internet use and skill for counties. The statistical models discussed in Table 2.2 also include a measure of the county density of domain name websites per 100 people (2018). This unique measure is constructed using data on 20 million domain name websites in the United States (see Mossberger et al. 2020). This de-identified and geo-coded data was shared with the authors by GoDaddy, the world's largest registrar of domain names. These are the underlying address book of the internet and an additional measure of internet use in geographic areas, especially for economic activity. According to random-sample surveys conducted by the company and shared with the authors, 75% of the domain name websites are commercial. We would expect both broadband adoption and the density of domain name websites to be positively related to measures of prosperity in counties. Additional analysis not shown here demonstrates that the results are essentially the same whether the variable measuring these more specific and largely commercial uses of the internet (the density of domain name websites) is included or omitted.

The results of the multivariate regression model shown in Table 2.2 suggest that broadband subscriptions and other measures of use (the density of website

Table 2.2 Predicting County Economic Prosperity Index with Broadband Subscriptions

	(1)	(2)	(3)
	Prosperity 2016	Change in Income 2016–2017 (Thousands)	Economic Recovery 2007–12/2016
Broadband Subscriptions	0.6258***	0.017292*	0.1065*
	(0.0843)	(0.00666)	(0.0409)
Density Domain Name Websites per 100	0.1383	−0.00336	0.5227***
	(0.1816)	(0.02497)	(0.1454)
Small Business Density per 100	0.2531**	−0.00624	−0.3282**
	(0.0909)	(0.01421)	(0.1195)
Percent Black	−0.1967***	−0.01035*	0.0306
	(0.0439)	(0.00482)	(0.0272)
Percent Native American	−0.2418***	−0.00266	0.0869
	(0.0658)	(0.00686)	(0.0476)
Percent Asian	0.1403	0.102***	0.0492
	(0.0874)	(0.02583)	(0.0778)
Percent Hispanic	0.0936*	−0.00689	0.0479
	(0.0426)	(.00534)	(0.0241)
Percent Agriculture	−0.4181***	−0.00365	−0.2904**
	(0.0932)	(0.02064)	(0.0903)
Percent Construction	1.0135***	.0735**	0.4573*
	(0.1819)	(0.02471)	(0.1950)
Percent Wholesale	0.0464	.021377	0.2101
	(0.4826)	(.08522)	(0.3090)
Percent Retail	−0.6597*	.00122	0.5434***
	(0.2555)	(0.01877)	(0.1141)
Percent Transportation	−0.1541	−0.02011	−0.6898***
	(0.2677)	(0.0170)	(0.1308)
Percent Information Technology	0.2348	.107851*	−0.1501
	(0.5312)	(0.0528)	(0.3704)
Percent Finance	−0.2141	−0.063*	−0.0919
	(0.1428)	(.02587)	(0.1764)
Percent Professional	−0.9216***	(.069855)	0.1929
	(0.2149)	(.03562)	(0.1811)
Percent Education	−0.4626**	−0.02385*	−0.4025***
	(0.1426)	(0.00938)	(0.1094)
Percent Other	−0.2389	−0.16211	−0.4636
	(0.4518)	(0.0856)	(0.3154)

Table 2.2 *Continued*

	(1)	(2)	(3)
	Prosperity 2016	Change in Income 2016–2017 (Thousands)	Economic Recovery 2007–12/2016
Percent Public	–0.4026**	–0.0158	–0.2889*
	(0.1467)	(0.0182)	(0.1232)
Percent High School	1.1820***	.001212	0.0710
	(0.1578)	(.01803)	(0.0734)
Percent College	0.5447***	.012043	–0.0879
	(0.0779)	(.00702)	(0.0790)
Percent Millennial	–0.9436***	–0.03933	0.0428
	(0.2028)	(.02582)	(0.1188)
Percent Gen X	0.5379	.007908	–0.3975*
	(0.3274)	(.02433)	(0.1930)
Percent Baby Boomer	–1.3432***	–0.02988	0.1550
	(0.2149)	(.02085)	(0.1303)
Constant	–50.5010**	2.420652	3.5794
	(17.9436)	(2.37312)	(10.4622)
Observations	2,958	2,958	2,958

domains) are indeed statistically significant predictors of overall economic prosperity across counties (column 1) and positive change in median income (column 2) but not change in economic prosperity (recovery from the recession: column 3). These strong positive relationships remain, controlling for the size of industry sectors, demographic factors, and educational attainment. More broadband subscriptions are statistically associated with the factors measured in the county economic prosperity index, including decreased poverty, higher educational attainment, more full-time employment, higher growth in median incomes,[7] and greater job and business growth.

We replicate the methodology used by the Economic Innovation Group and extend the statistical models to estimate economic prosperity in 2018, as well as change in median income from 2016 to 2018 using updated data from the ACS. We again find that broadband subscriptions in the population have a large, positive, and significant relationship with economic prosperity in 2018. A 1% increase in the broadband subscription rate increases prosperity by .36; thus, a 10% increase in connectivity would lead to a 3.6-point growth in prosperity, all else equal. Broadband is unrelated to changes in prosperity from 2007 to 2018 but does have a strong positive relationship with growth in household median income from 2016 to 2018. A $40 increase in income for every 1% increase in broadband subscription is substantively a very large finding given that the total change in income in this period is $3,600. So, a community increasing broadband subscription rates by 10% would see incomes grow over 10% faster compared to the nation. These results show that the positive effects of broadband hold even when extending models to new time periods.

What is the substantive size of this relationship for broadband adoption? Predicted values (Figure 2.6), or simulations based on the regression coefficients from Table 2.3, with other factors held constant, show that overall economic prosperity increases 30 points when moving from a county with 20% of the households with home broadband to 80%, with all other variables held constant at mean values. It also shows that the average change in income from 2016 to 2018 increases from $2,600 to over $5,000 as a community goes from 20% to over 80% broadband subscription rates.

But is some factor driving both economic outcomes across counties and broadband connectivity rates? Using instrumental variable causal models, we apply a stronger test, stripping out the effects of factors that might influence broadband subscriptions in the first place: small business density and the percent rural population. Counties with more small businesses may have a higher need for broadband than places with less economic activity. Rural areas are known to have lower rates of broadband adoption than metropolitan regions (Whitacre et al. 2014a; Mossberger et al. 2013). Even after using these two instrumental variables to remove from our measure of broadband the variation it shares with measures of small business density and rural populations, broadband subscriptions are a strong positive predictor of community economic prosperity, further suggesting causation. The effect is substantively large as well, as shown in Table 2.4 and Figure 2.7. As broadband subscription in counties increases, so do economic prosperity and median income, even using the instrumental variable causal model.

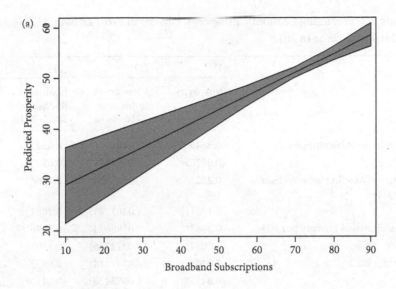

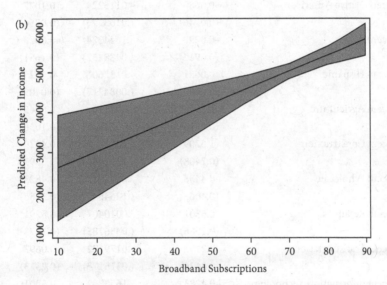

Figure 2.6 Predicted county economic prosperity and change in median income, 2016–2018, varying broadband subscriptions.

Table 2.3 Predicting Economic Prosperity, Change in Prosperity, and Change in Median Income in 2018

	(1)	(2)	(3)
	Prosperity 2018	Change in Income 2016–2018	Economic Recovery 2007–2018
Broadband Subscriptions	0.3645***	.0404967***	0.0327
	(0.0620)	(.095181)	(0.0722)
Density Domain Name Websites per 100	0.2523	.0418974	0.7866***
	(0.1721)	(.0407721)	(0.2081)
Small Business Density per 100	−0.2863**	−.0198624	−0.7476***
	(0.1054)	(.0212246)	(0.1593)
Percent Black	−0.0203	−.0249114**	0.0692
	(0.0455)	(.0083420)	(0.0416)
Percent Native American	−0.0680	−.0115229	0.1077
	(0.0746)	(.0103775)	(0.0754)
Percent Asian	−0.0298	.1644274**	−0.0523
	(0.0912)	(.0488452)	(0.1651)
Percent Hispanic	0.0681	−.0132606	−0.0125
	(0.0390)	(.0084217)	(0.0409)
Percent Agriculture	0.2546**	−.002216	0.0845
	(0.0811)	(.0324718)	(0.1058)
Percent Construction	1.5209***	.1371888***	0.4429
	(0.2408)	(.0351068)	(0.3597)
Percent Wholesale	−0.4438	.0045011	−0.4823
	(0.4065)	(.1264545)	(0.5636)
Percent Retail	0.5301*	.0010407	0.8291***
	(0.2586)	(.0336785)	(0.1773)
Percent Transportation	0.2193	−.0179172	−0.0697
	(0.2626)	(.0315173)	(0.2323)
Percent Information Technology	−0.4283	.1658722	0.5901
	(0.4894)	(.0973627)	(0.6467)
Percent Finance	0.0466	−.0906136*	−0.2673
	(0.1272)	(.0402266)	(0.1763)
Percent Professional	−0.4285	.0688828	0.0344
	(0.2806)	(.0657809)	(0.2741)
Percent Education	−0.8071***	−.0459723**	−0.6404***
	(0.1041)	(.0140847)	(0.1728)
Percent Other	−0.4485	−.2979459*	−0.9080*
	(0.4061)	(.1188523)	(0.4309)

Table 2.3 *Continued*

	(1)	(2)	(3)
	Prosperity 2018	Change in Income 2016–2018	Economic Recovery 2007–2018
Percent Public	−0.3251*	−.0290270	−0.3907
	(0.1270)	(.0261200)	(0.2056)
Percent High School	1.2952***	−.0181361	−0.0052
	(0.1288)	(.0286928)	(0.1174)
Percent College	0.5177***	.0392767**	−0.0609
	(0.1056)	(.0137109)	(0.1096)
Percent Millennial	−0.3479	−.0465137	0.3583
	(0.2375)	(.0460212)	(0.2300)
Percent Gen X	0.3846	.0531534	−0.1075
	(0.2446)	(.0399580)	(0.3138)
Percent Baby Boomer	−0.4652	−.0293028	0.5809*
	(0.2609)	(.0441067)	(0.2382)
Constant	−82.2087***	3.8647534	−4.5303
	(19.9522)	(4.33637)	(21.3717)
Observations	2,934	2,958	2,894

Does Broadband Benefit All County Types Equally: Urban Core, Suburban, Rural? Interaction Models

Does broadband improve economic outcomes for some communities more than others? The Forman et al. (2012) study on broadband use by firms discovered that it produced spillover benefits only for the largest metropolitan counties. Whitacre et al. (2014a) examined broadband adoption only in non-metropolitan counties. Using data from the Pew Research Center and the Centers for Disease Control and Prevention, we identify urban core counties with over 1 million people (n = 63), suburban counties and counties with smaller cities (n = 1,050), and rural counties (n = 1,838) to explore heterogeneity across counties in our models. This is a conservative measure for urban areas, measuring only large urban core counties, but these are precisely the counties identified as most likely beneficiaries of broadband business use in Forman et al. (2012). We create an indicator variable measuring whether a county is urban, suburban, or rural and multiply it by the percentage of the population with broadband subscriptions. This interaction term in our

Table 2.4 Instrumental Variable Models: Predicting County Prosperity with Broadband Subscriptions

	(1)	(2)	(3)
	Economic Prosperity	Median Income (Thousands $)	Economic Recovery
Broadband Subscriptions	0.2814***	.3875748***	−0.1307
	(0.1023)	(.1051264)	(0.1168)
Percent Black	−0.0329	−.1343598***	0.0424
	(0.0306)	(.0288889)	(0.0399)
Percent Native American	−0.0857	−.1718712***	0.0864
	(0.0758)	(.0517401)	(0.0785)
Percent Asian	0.0057	.6129559***	0.0157
	(0.0835)	(.0905885)	(0.0864)
Perc Hispanic	0.0858**	−.0251755	0.0338
	(0.0334)	(.0245742)	(0.0342)
Percent Agriculture	0.1750**	.3201191***	−0.1216
	(0.0723)	(.0634824)	(0.0854)
Percent Construction	1.4749***	.9793524***	0.3019
	(0.1888)	(.1522574)	(0.2883)
Percent Wholesale	−0.4619	.2883769	−0.6018
	(0.3752)	(.2849059)	(0.4465)
Percent Retail	0.5199***	−.7169846***	0.7797***
	(0.1560)	(.1467795)	(0.1809)
Percent Transportation	0.2191	.1671872	−0.0763
	(0.1804)	(.1470181)	(0.1956)
Percent Information Technology	−0.5589	−.0655327	0.3341
	(0.5010)	(.4438615)	(0.6760)
Percent Finance	0.0409	−.0799272	−0.2640
	(0.1707)	(.1937665)	(0.2260)
Percent Professional	−0.2755	−.0921564	0.4392**
	(0.1750)	(.1629381)	(0.1997)
Percent Education	−0.7933***	−.0213398	−0.6378***
	(0.0811)	(.0866961)	(0.0823)
Percent Other	−0.4546	−1.0988031***	−0.9329***
	(0.3461)	(.3173274)	(0.3521)
Percent Public Sector	−0.2871***	.9275522***	−0.3292**
	(0.1081)	(.1253803)	(0.1629)
Percent High School	1.3596***	.1766431*	0.1190
	(0.1267)	(.1007093)	(0.1294)

Table 2.4 *Continued*

	(1)	(2)	(3)
	Economic Prosperity	Median Income (Thousands $)	Economic Recovery
Percent College	0.4767***	.7531172***	−0.1678**
	(0.0647)	(.0573534)	(0.0792)
Percent Millennial	−0.3718**	−1.7777909***	0.3417*
	(0.1593)	(.1593048)	(0.1813)
Percent Gen X	0.4517**	.7659976***	0.0608
	(0.1954)	(.1805697)	(0.2104)
Percent Baby Boomer	−0.5491***	−1.1395239***	0.4181**
	(0.1588)	(.1365103)	(0.1689)
Constant	−84.5368***	34.7952265***	−10.5851
	(14.5786)	(12.2129039)	(14.9529)
Observations	2934	2958	2894

multivariate regression model tests for differential effects from broadband subscriptions for different places. The same set of covariates used in this analysis is reported in the previous regression models.

The results of the interaction models demonstrate that broadband adoption increases economic prosperity for all types of counties (urban, rural, and suburban) but has a particularly strong effect in suburban counties. Figure 2.8 (based on the regression coefficients, with all other variables held constant) shows that suburban counties go from least to most prosperous when broadband access increases, but the benefits of broadband are experienced across county types.

While urban, rural, and suburban counties all experienced higher median income when broadband connectivity increased, again, we find suburban areas benefit the most. Some studies of broadband deployment and the attraction of knowledge-intensive firms suggest that broadband infrastructure improves outcomes most for communities on the metropolitan fringe or in smaller communities that are strategically located between major metropolitan areas (Mack 2014). These results indicate that broadband adoption also has varied effects on different types of communities. Initial analysis suggests that suburban areas benefit the most from increased broadband adoption, consistent with the earlier metropolitan fringe finding. Broadband use in the

(a)

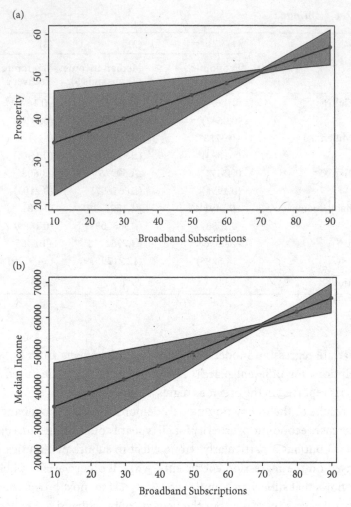

(b)

Figure 2.7 Instrumental variable models: predicted county economic prosperity and median income, varying broadband subscriptions.

population may also make these areas more attractive for investment, generating benefits for communities.

But because these data are measured at just one time period (2016–2018), this relationship may be correlational, not causal. The previous analysis doesn't address trends over time, control for duration (time series trends), or lag the explanatory variable in predicting economic prosperity. Time series data is needed.

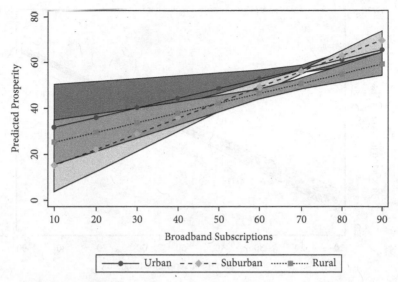

Figure 2.8 Predicted county economic prosperity, varying broadband subscriptions, in urban, rural, and suburban counties.

Time Series Analysis: Predicting County Economic Outcomes over Time

We use pooled cross-sectional time series data for counties from 2007 to 2017 to predict the effect of the percentage of the population with broadband on county median income (measured in thousands of dollars). Our sample includes the largest population counties—324 observations per year—where data is available over time. Median income is used because the prosperity index is not available except in two time periods. An alternative specification includes the 324 observations for the early time period (2007–2012), 800 counties for 2013–2016, and all 3,100 observations for 2017 to create a total sample of 5,400 observations over the 10-year period.

The statistical models include demographic controls for educational attainment and racial and ethnic group size. Additionally, they control for county population (as larger, more populated counties have been shown to have better outcomes in studies of broadband use in business firms) (Forman et al. 2012). All covariates are lagged by one year for directionality—predictor variables should be measured in the year prior to the outcome (Figure 2.9). The models include fixed effects for year to control for the upward trend in broadband connectivity over time and by state to test whether broadband

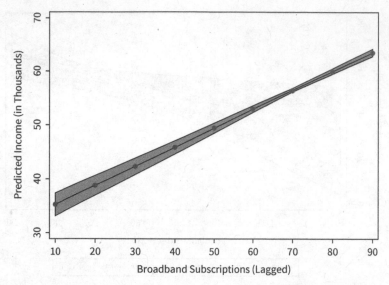

Figure 2.9 Predicted county median income, varying lagged broadband subscriptions (time series model).

matters more in some states than others. The results reported in Tables 2.5 and 2.6 indicate that broadband subscriptions remain a strong positive predictor of improved economic outcomes using the time series data. Controlling for other factors, a 1% increase in broadband subscriptions is associated with a $350 increase in median income for the largest 324 US counties over time, and an even larger $430 increase when evaluating all counties.

Each of the statistical analyses presented so far in this chapter points to the development of a digital economy where technology and human capital combine to spur economic growth. We can measure different dimensions of human capital in the economy by interacting broadband subscriptions and college-educated populations to predict economic prosperity. We ask whether broadband has the same effect on prosperity in counties with different levels of urbanism (percent rural), racial and ethnic diversity (percent white non-Hispanic), educational attainment (percent college degree), percentage of the population working in the IT industry, or young people (percent millennial age cohort). We divide the sample of 3,100 counties from the contemporary time period (2017) into quartiles by these four characteristics and then interact the quartile measure by broadband subscriptions. Of these many statistical tests, the only interaction that shows a statistical significance with broadband subscriptions is education. As shown in Figure 2.10,

Table 2.5 Time Series Model Predicting Median Income Using Lagged
Broadband Subscriptions, Counties 2007–2017: Largest 324 Counties

	(1)	(2)
	Income	Income
Lagged Broadband	0.3537*	0.0478
	(0.0162)	(0.0298)
Broadband		0.4337*
		(0.0358)
Population	0.0000	0.0000
	(0.0000)	(0.0000)
Percent Black	–0.3907*	–0.3715*
	(0.0186)	(0.0183)
Percent Native American	–0.1571*	–0.0734
	(0.0687)	(0.0676)
Percent Asian	1.1238*	1.0614*
	(0.0460)	(0.0453)
Percent Hispanic	–0.1716*	–0.1457*
	(0.0158)	(0.0156)
Percent College	0.3882*	0.3597*
	(0.0285)	(0.0279)
Percent High School	–0.1069*	–0.1272*
	(0.0089)	(0.0089)
Constant	35.1923*	24.0723*
	(3.6276)	(3.6641)
Observations	3,213	3,212

counties in the top quartile with the highest percentage of college graduates
or above benefit the most from widespread broadband adoption and experience the strongest economic prosperity. This may be the most important
finding of this chapter—education and technology access combine to create
economic opportunity in the digital age.

Broadband Availability versus Broadband Subscriptions

In a highly cited article, Whitacre et al. (2014a) compare the effects of both
broadband availability (infrastructure) and adoption (subscriptions) for

Table 2.6 Time Series Model Predicting Median Income Using Lagged Broadband Subscriptions: Full Sample of Counties

	(1)	(2)
	Income	Income
Lagged Broadband	0.4316*	0.0711*
	(0.0146)	(0.0253)
Broadband		0.5112*
		(0.0297)
Population	0.0000*	0.0000
	(0.0000)	(0.0000)
Percent Black	−0.2708*	−0.2522*
	(0.0146)	(0.0143)
Percent Native American	−0.0641	0.0161
	(0.0416)	(0.0408)
Percent Asian	1.1067*	1.0183*
	(0.0427)	(0.0418)
Percent Hispanic	−0.1013*	−0.0789*
	(0.0138)	(0.0135)
Percent College	0.3760*	0.3364*
	(0.0211)	(0.0207)
Percent High School	−0.0701*	−0.1004*
	(0.0059)	(0.0060)
Constant	27.5459*	15.3648*
	(2.6797)	(2.7007)
Observations	5168	5166

economic growth in rural counties using FCC data from 2000 to 2010. They find the impact of availability is limited and adoption matters more, for income and unemployment. We expand their analysis using longer time series data from rural, urban, and suburban counties and our measure of broadband subscriptions.

We compare the percentage of the population with a broadband subscription using census data (measuring both fixed and mobile) to comparable data measuring fixed broadband deployment (availability of service), or the number of service providers from the FCC, building on Whitacre et al. (2014a). The subscription data from the ACS is also more precise, as the FCC subscription data is available only for quintiles. Our hypothesis is that actual

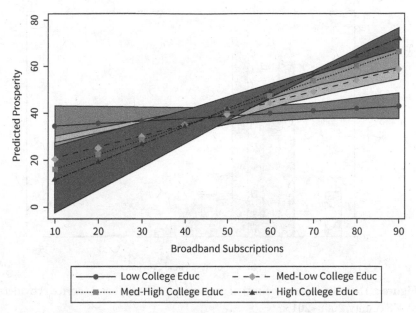

Figure 2.10 Predicted county economic prosperity, varying broadband subscriptions for counties by college-educated population.

broadband use as measured by broadband subscriptions should have a larger effect on economic outcomes (median income growth) than the availability of broadband infrastructure. That is, adoption and use are what matters most.

Figure 2.11 shows the distribution (histogram) for the number of service providers across the 3,000+ US counties. The distribution is surprisingly normal with most counties having between five and seven service providers.

To conduct the replication analysis, we restrict the sample of counties to those for which we have data on broadband subscriptions from the census (the largest-population counties). The FCC's measure of the number of providers for fixed broadband is only available through 2014, so we truncate our time series to 2014 to perform this replication (Table 2.7, columns 1 and 2). To measure change in median income, a lagged panel model is used, including median income in the county in the prior year as a predictor. This means the outcome variable is change in median income.

The results are reported in Table 2.7. Variables that are statistically significant in predicting annual change in median income (in thousands of dollars) are noted by an asterisk. The percentage of broadband subscriptions predicts change in median income over time, but the FCC's fixed broadband

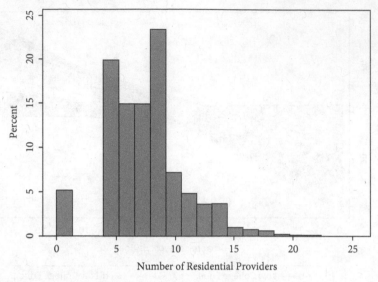

Figure 2.11 Distribution (count) of the number of broadband service providers by US county, 2008–2013.

deployment measured by the number of providers does not (see column 2). This is strong evidence that it is broadband use that matters for economic prosperity.

Next, instead of using our measure of the percentage of the population with a broadband subscription, in column 1 we measure fixed broadband subscriptions using the FCC data. This is an ordinal variable ranging from 0 to 5, reflecting how much of the counties' population is online: 0 = 0%, 1 = 0–20%, 2 = 20–40%, 3 = 40–60%, 4 = 60–80%, and 5 = 80–100%. Only counties with 80% of the population with a fixed broadband connection or higher have statistically higher growth in median income compared to those with between 0% and 20%.

Column 3 and 4 replicate the models in columns 1 and 2 but include the data for the entire time series, omitting the variable for fixed broadband deployment (availability). The same results hold.

There is strong evidence that census data on broadband subscriptions is a significant improvement over measures of broadband availability for predicting outcomes, as shown in Figure 2.12. Holding other predictors of the growth of median income constant, including demographic factors and educational attainment, as the percentage of the population with broadband

Table 2.7 Predicting Change in Median Household Income across US Counties, 2008–2014 (Columns 1 and 2) and 2007–2017 (Columns 3 and 4)

	(1)	(2)	(3)	(4)
	Change in Income	Change in Income	Change in Income	Change in Income
Lag Income	0.9734*	0.9722*	0.9875*	1.0073*
	(0.0021)	(0.0040)	(0.0014)	(0.0019)
Lag Broadband Subscriptions (%)		0.0485*		0.0117*
		(0.0040)		(0.0016)
FCC (20–40% Broadband)	0.1662		0.0750	
	(0.1349)		(0.1226)	
FCC (40–60% Broadband)	0.0918		−0.0555	
	(0.1333)		(0.1199)	
FCC (60–80% Broadband)	0.2352		0.0392	
	(0.1377)		(0.1213)	
FCC (80–100% Broadband)	0.5698*		0.4426*	
	(0.1529)		(0.1284)	
Lag Number of Residential Providers	−0.0310*	−0.0225		
	(0.0070)	(0.0148)		
Lag Population	−0.0000	−0.0000	−0.0000	−0.0000
	(0.0000)	(0.0000)	(0.0000)	(0.0000)
Lag Percent Black	−0.0105*	−0.0076	−0.0088*	−0.0034
	(0.0017)	(0.0040)	(0.0012)	(0.0019)
Lag Percent Native American	−0.0048	0.0098	−0.0026	0.0047
	(0.0026)	(0.0104)	(0.0020)	(0.0038)
Lag Percent Asian	0.0356*	0.0242*	0.0345*	0.0158*
	(0.0095)	(0.0117)	(0.0066)	(0.0069)
Lag Percent Hispanic	−0.0028	0.0039	−0.0005	0.0019
	(0.0019)	(0.0037)	(0.0012)	(0.0019)
Lag Percent College	−0.0122	−0.0397*	0.0368*	0.0370*
	(0.0077)	(0.0158)	(0.0024)	(0.0026)
Lag Percent High School	−0.0071	−0.0030	0.0193*	0.0277*
	(0.0038)	(0.0087)	(0.0005)	(0.0007)
Constant	3.6914*	0.5029	0.7111*	−2.5625*
	(0.3724)	(0.8952)	(0.2125)	(0.3082)
Observations	16351	2419	28405	7977

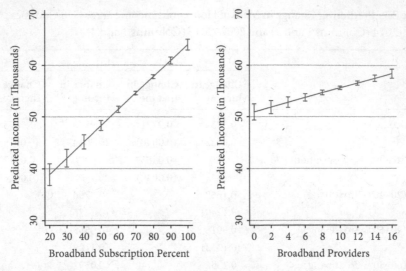

Figure 2.12 Predicted median household income (in thousands) as broadband subscriptions percent (left) or broadband availability (number of providers) increases, all else equal.

rises from 60% to 90% of the population, median household income climbs from $50,000 to over $60,000.

Similarly, as shown in Figure 2.13, our interval measure of broadband subscriptions provides much more explanatory power in predicting higher median income growth than the FCC's quintile measure, which only shows a significant increase in household income for counties with 80% of the population online (or higher). By replicating the previous analysis conducted by Whitacre et al. (2014a) with the more precise ACS data, we can see the additional power this conveys over the quintile measures, as well as how broadband adoption matters for all types of counties.

Conclusion: Broadband Impacts in Diverse Counties

As inequality has grown across places, local policymakers in both rural and urban communities have viewed technology as a resource to address disparities and promote local development. Counties play an essential part in regional economic development and human service delivery, and so are concerned with creating regional opportunity. This chapter presents the first

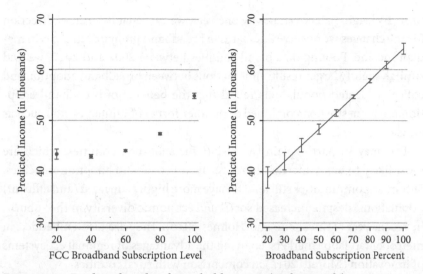

Figure 2.13 Predicted median household income (in thousands) using different sources for broadband subscription rates (Federal Communications Commission (FCC) quintiles on left, ACS broadband data on right)

evidence that digital human capital contributes to prosperity in both metropolitan and nonmetropolitan counties. Prior research has focused on broadband deployment rather than adoption and use, except for some research on nonmetropolitan counties.

Employing new census data on broadband adoption, we are able for the first time to examine precise estimates on subscriptions for all counties, rather than data that is limited to only the largest counties or to quintiles. Using this 2017 data, we find that broadband adoption is strongly related to higher levels of community prosperity, median income, and recovery from the recession through 2016. While data for all counties is available only for a single year, the recovery variable is a measure of change over two time periods and provides more persuasive evidence for broadband adoption as one cause of that change. The instrumental variable causal models further strengthen the case for broadband's impact.

Time series analysis offers the most convincing evidence that broadband adoption influences community outcomes. With historical estimates described in Chapter 1 for the 324 largest counties, we find that between 2007 and 2017, a 1% increase in broadband adoption produced on average a $350 increase in median income, controlling for other factors. The prosperity index, which takes a broader set of impacts into account, is not available over time. By lagging all covariates by one year, we can establish causal direction for the changes we observe: increases in broadband prefigure changes in median income. Pooling data on all counties between 2007 and 2017, we find similar, slightly larger results. Interactions between broadband adoption and college-educated populations reveal that the benefits of broadband adoption are highest when combined with other forms of human capital such as education.

This may in part explain the results for suburban counties, which are most likely to enjoy increased prosperity as broadband adoption increases. Suburban communities still tend to have more highly educated (and affluent) populations, despite increased social and economic diversity in the suburbs in recent years (Kneebone and Holmes 2016). They are also embedded in metropolitan labor markets, enjoying the advantages of regional ecosystems of innovation (Moretti 2012) in comparison with rural counties.

The "suburban" category used here also included smaller metros outside the largest urban counties. Smaller tech hubs have begun to flourish outside of the major cities, taking advantage of lower costs of living, different talent pools, and geographic diversity. Start-ups increasingly locate in college towns and in smaller metros in the interior of the country (including Pittsburgh, Pennsylvania; Columbus, Ohio; Des Moines, Iowa; and Omaha, Nebraska) (Florida 2019). More generally, smaller metros and micropolitan areas play an important role in economic development in rural areas (Sisson 2018). In the next chapter, we explore outcomes for the 50 largest metropolitan areas (both urban core and suburban communities), including the superstar city regions and places that have relied on manufacturing for their economic lifeblood.

Finally, this chapter examined whether the more precise data on broadband subscriptions over time that we use in this book is a better measure of broadband adoption than the FCC quintile data, and a stronger predictor of broadband's impacts in comparison with broadband availability (the number of providers). We replicate the earlier Whitacre et al. (2014a) research that compared availability with the quintile measures of broadband adoption. Like these authors, we found that broadband subscriptions are more likely to predict positive economic outcomes for communities than availability, and that these new measures show impacts for median income at lower levels of adoption as well as more pronounced effects overall. In this replication, as well as throughout the chapter, we have explored outcomes for all counties: urban, suburban, and rural.

Surveying the effects of broadband use across counties provides evidence that widespread adoption increases prosperity in all types of counties. Suburban counties like Douglas County, Colorado, with their high rates of resident connectivity, can produce award-winning innovations in digital government. But Apache County, Arizona, and Cuyahoga County, Ohio, can realize important gains too. Broadband adoption represents a form of digital human capital for inclusive growth in diverse contexts.

3

Metros

Does Broadband Promote Growing and Prosperous Regions?

With Kellen Gracey

When Amazon announced its decision to create two "second headquarters," policy analysts weighed in that the choices of the New York and Washington metros were entirely predictable (Barker and Lanham 2018). Both places were situated on the East Coast rather than the heartland, with well-educated workers, population density, and information technology (IT) talent (Liu 2019; Berube 2018) that rendered them obvious picks. The decision followed a lengthy public competition with two rounds and 238 applicants vying for the projected 50,000 jobs. The protracted process allowed Amazon to gather data on incentives and programs in cities around the country, as well as to extract substantial inducements in the high-stakes bidding (Streitfeld 2018; Jensen 2019). While some cities focused more on marketing their economic potential and existing incentive programs, Dallas offered a 99-year package, North Carolina passed new legislation to increase tax rebates for Amazon, and Kankakee, Illinois, suggested that Amazon could create a bank so that it would be eligible for incentives for financial institutions (Jensen 2019). Such a free-for-all encourages proposals with questionable consequences for the public interest and a need for alternative strategies. A political backlash against the $3.5 billion offered by New York City prompted Amazon to reverse its decision to split the headquarters. It also provoked local and national debates about subsidies to big tech firms (Goodman 2019; Jensen 2019).

Research and policy discussions on technology's economic impacts in metropolitan areas have focused on IT employment and investment, and its concentration in tech-driven "superstar cities" (Florida 2017; Moretti 2012; Badger 2018; Milken Institute 2018). This is an important dimension of current economic trends, but a focus on big tech firms alone neglects other aspects of the economy in the digital age. We ask how the connectivity of

Choosing the Future. Karen Mossberger, Caroline J. Tolbert, and Scott J. LaCombe, Oxford University Press. © Oxford University Press 2021. DOI: 10.1093/oso/9780197585757.003.0003

metropolitan populations affects regional outcomes, across occupations and sectors, as technology use becomes increasingly important for a range of jobs and economic activities. Do the skills and information enabled by technology use contribute a form of digital human capital in the workforce that is broader and more inclusive than the high-tech jobs or superstar cities?

We examine evidence on outcomes for broadband adoption in the 50 largest metros between 2000 and 2017. Because of their size and density, these are precisely the places that have high potential for innovation, though there are wide disparities across these regions.

Metros are the connective tissue of the economy and centers of opportunity (Shearer et al. 2018). As labor markets defined by commuting patterns, they represent the majority of the nation's economic activity. The top 100 metropolitan regions boast two-thirds of the population and 75% of the national economy (Katz and Bradley 2013, 1). Metros are diverse, and in a large nation like the United States, there is not a single economy, but a network of metropolitan economies (Katz and Bradley 2013, 1). One of the ways in which metro areas differ is in the connectedness of their populations.

Urban disparities in broadband adoption are starkest in low-income neighborhoods (Tomer et al. 2017), but differences are visible at the metropolitan level too. For the 50 largest metros, where economic and social activity increasingly rely upon technology, the percentage of the population with broadband subscriptions ranged from a little less than 71% in Memphis, Tennessee-Mississippi-Arkansas, to more than 89% in San Jose–Sunnyvale–Santa Clara, California, in 2017 (ACS five-year estimates). Both metros are clearly more dependent on technology use than in 2000, when the Memphis area had less than 5% of the population with broadband at home and the San Jose conurbation had just under 13%, according to the data we analyze here. But the gaps have persisted and in fact widened in the ensuing decades.

How do differences in broadband use over time affect economic opportunity across metropolitan regions? How do they matter for outcomes such as community prosperity and economic growth? Drawing on large-sample census data, we use multilevel statistical models and poststratification weights to estimate broadband subscriptions for the population and its impact on full-time employment and the Brookings Metro Monitor measures of growth and prosperity. In this chapter, we can examine outcomes over a longer time period than in Chapter 2, with nearly two decades of data on broadband subscriptions in the nation's 50 largest metropolitan areas from 2000 to 2017.

We examine economic impacts in the nation's largest regional economies at a time when inequality across places is increasing, and we consider the implications of broadband adoption for inclusive metropolitan development and economic opportunity.

Superstar Metros and Rising Inequality

Cities and metropolitan areas have been heralded as the drivers of innovation in the economy and society, both in the United States and around the world (Katz and Bradley 2013; Glaeser 2011; Katz and Nowak 2017; Bouchet et al. 2018). Writing about the "metropolitan revolution" in 2013, Brookings authors Katz and Bradley called metros the "engines of economic prosperity and social transformation" (Katz and Bradley 2013, 1). Not all metropolitan areas, however, have shared equally in this revolution—a reality that has become clearer with the passage of time. The concentration of investment and wealth on the coasts and in other technology hubs, compared with the lagging fortunes of other metropolitan regions, has prompted concerns about rising inequality across places and its political and social consequences (Berube and Murray 2018; Florida 2017; Moretti 2012).

Between 2010 and 2017, nearly half of the nation's employment growth occurred in just 20 metropolitan areas, and between 2007 and 2016, the New York and Washington metros accounted for about half of the nation's increase in business establishments (Badger 2018). Metros such as San Francisco, San Jose, Austin, Raleigh-Durham, and Boston-Cambridge, which have the highest share of high-tech jobs and venture capital investment, are flourishing, while the older industrial metros of the heartland have struggled (Moretti 2012, 84–85; Berube and Murray 2018). According to Moretti, this divergence is deepening and accelerating (Moretti 2012, 14). Though COVID-19 has enabled telework and prompted concerns about densely populated cities, there may be no long-term or fundamental reversal of these trends (Muro 2020; Florida and Pedigo 2020). Agglomeration has increased with technological change and greater dependence on human capital, privileging dense urban areas with highly educated workers and knowledge spillovers that resonate throughout the economy (Glaeser 2011, 6). While there are greater inequalities across places, workers from all occupations benefit from the collective human capital in innovation hubs; positive externalities or spillover effects mean that wages are higher across industries, including for less educated and less skilled workers (Moretti 2012, 89, 100).

As with other forms of human capital, we argue that the skills and information acquired by individuals through broadband adoption could be expected to, in the aggregate, create regional spillovers and multiplier effects. In contrast with the share of IT jobs or other measures of the digital economy, broadband adoption (percentage of the population with subscriptions) captures digital human capital throughout the metropolitan labor market, across sectors, for the highly skilled, and for less educated workers.

We are also interested in whether a higher presence of younger workers, who have grown up in the age of the internet, might signal additional digital human capital in the region. A recent analysis by Brookings that identifies the top cities for educated millennials includes centers of technology investment and employment. The 15 metropolitan areas with the highest shares of millennials, according to Brookings, "are all in the fast-growing South and West, such as Austin, San Diego, and Los Angeles. The lowest millennial shares tend to be in Florida, such as Tampa and Miami, in the Northeast, such as Pittsburgh, and in the Midwest, such as Cleveland and Detroit" (Frey 2018).

Does digital human capital—the ability to access and use IT throughout the community (Bach et al. 2013)—play a role in metropolitan economic outcomes? Inclusive metropolitan growth is based on the premise that such regions can draw upon a broader range of skills and deeper pools of talent (Florida 2017; Parilla 2017). If more inclusive metros, defined as having a larger share of their population with connectivity, have more widespread digital human capital, then broadband adoption should be related to greater regional economic benefits.

Broadband and Economic Outcomes

This study extends prior research on broadband's impacts on economic development in several important ways: it examines broadband use rather than infrastructure and population use rather than business applications, testing the proposition that digital human capital creates local economic opportunity. It focuses on metropolitan areas, which are the nation's most productive regional economies.

Overall, the literature has pointed to positive effects for broadband deployment (see Holt and Jamison 2009; Falck 2017; Bertschek et al. 2016; and Abrardi and Cambini 2019 for international overviews). Some research indicates that broadband availability increases employment and growth more generally, but especially in IT-intensive firms (Lehr et al. 2006). Kolko's national study (2012) found an association between broadband deployment and economic growth, though this was most likely in industries that employ technology intensively and in places with less population density rather than metros. Kolko concluded there was no relationship between deployment and local employment rates or wages, possibly because existing residents did not get jobs being created. In a similar vein, a Swedish study revealed that broadband deployment and use by firms increased the productivity and wages of skilled, college-educated workers but substituted for unskilled workers who performed routine tasks, and therefore widened wage inequality (Akerman et al. 2015).

Internet use by firms in the United States has been shown to lead to greater wage growth in metropolitan regions than in rural areas, because metropolitan firms are able to employ technology in more advanced and complex ways (Forman et al. 2005, 2008).

This growth, however, does not appear to be equitable across metropolitan regions. With firm-level data, Forman et al. (2012) demonstrated that advanced use of the internet (beyond email and basic search) enhances productivity and is related to wage and employment growth, but only in 6% of counties, which have large populations, high incomes, and high IT industry presence. Advanced internet use leads to positive outcomes for some, but also increases regional wage inequality (Forman et al. 2012). This is consistent with the tale of regional divergence and rising inequality (Moretti 2012; Florida 2017), indicating that concentration of the IT industry may be a driving force.

Deployment and business use of high-speed internet are certainly relevant to the questions we ask, but neither captures all the ways in which internet access and skills could benefit residents and create local spillovers, including uses beyond jobs requiring the most education. In regions with higher levels of broadband adoption and digital skills, there may be more opportunities for workers to employ technology with different levels of education, and local workers may better compete for new jobs being created. Population use of technology is the missing element in this prior literature.

Missing Pieces of the Puzzle: Broadband
Subscriptions over Two Decades

We use new broadband subscription data over time to predict economic outcomes for metros, including (1) full-time employment measured by the census and (2) measures of growth and prosperity using the Brookings Institution's Metro Monitor data (described under data and methods later). Studying trends in broadband subscriptions allows us to understand the extent to which the technology is being used by the population and to assess whether this has resulted in meaningful change.

Our longitudinal analysis begins in 2000, the first year census data on broadband is available. Since 2013, the 3-million-person American Community Survey (ACS) has provided data for metros on the percentage of the population with a broadband subscription. To fill in the gaps prior to 2013, we estimate broadband subscription data for the 50 largest metros using the 100,000-person sample from the Current Population Survey (CPS) for the years 2000 (the first year available) to 2012. The same question wording is used in both the ACS and CPS. Cell phones and satellite are included as broadband in addition to cable, fiber, and DSL, but dial-up is not.

To generalize from the small survey sample for many of the metros can be problematic and lead to bias, so the common approach of simple disaggregation is not used for data from 2000 to 2012. To overcome this problem, we use the US Census Bureau's CPS microdata merged with metro-level variables, and multilevel statistical modeling with poststratification weighting to estimate broadband subscriptions (adoption) for metros for each year of the time series (see Mossberger et al. 2013 for a similar method). Multilevel regression with imputation and poststratification weights to develop population estimates of internet access for geographic areas is more precise and robust than survey disaggregation, leading to less error and narrower confidence intervals in our predictions (Lax and Phillips 2009a, 2009b; Park et al. 2004, 2006; Pacheco 2011). The result is time series data for broadband subscriptions for metros that was previously unavailable.

The model in Figure 3.1 illustrates the multilevel structure of the data estimating metro broadband adoption. The dependent variable y (broadband subscription) is a function of i characteristics. Each variable is modeled as a draw from a normal distribution with a mean of zero and an estimated variance. The respondents are nested within metros and states, and the metro estimates are also modeled as a function of metro-level poverty rates and

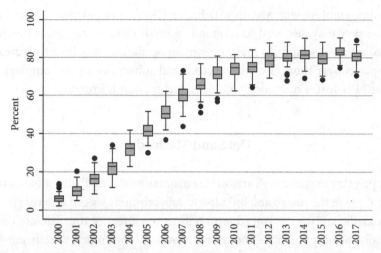

Figure 3.1 Broadband subscriptions of any type (including cell, percentage), 50 largest US metros, 2000–2017.

percent Black population. Individual-level variables measuring demographic factors (race/ethnicity, gender, education, income, age) and economic factors (North American Industry Classification System [NAICS] occupational categories) are merged with aggregate-level variables for metropolitan regions. With this multilevel data (individuals nested in communities) and the statistical method using probability simulations, we generate new estimates for broadband subscriptions for metropolitan areas over time. Metro-level estimates for each year are generated separately and then combined to create the time series for the years 2000–2012. These data are then combined with metro data measuring broadband subscriptions available from the US Census Bureau's ACS from 2013 to 2017. Together the time series spans 17 years beginning in 2000.

Research Questions

Does increased broadband connectivity promote prosperous regions? Do metropolitan areas with a higher share of broadband subscriptions in the population over nearly two decades have more economic prosperity, growth, and employment? We hypothesize that changes in broadband subscriptions over time are associated with increased economic prosperity, growth, and full-time employment. Metros with high millennial populations that are digitally savvy may also lead to economic growth in some regions. Growth in broadband subscriptions and millennial population may lead to increased prosperity over time. Growth in broadband subscriptions and employment in the IT industry may also combine to fuel economic growth.

Data and Methods

Our primary explanatory variables are a measure of broadband subscriptions (any type) in the metro and broadband subscriptions lagged one year ($t-1$). The coefficient for the lagged term tells us the effect of the absolute rate of broadband adoption on economic outcomes. By combining the current-year rate of broadband subscription with broadband in the prior year, our explanatory variable is effectively *the change* in broadband adoption each year. This creates a high bar for understanding whether broadband results in changes in economic outcomes.

We use three outcome variables in this analysis. All three measures work hand in hand to paint a picture of the economic landscape in America's top 50 metros. For two of the outcome variables we turn to the Brookings Metro Monitor (Shearer et al. 2018), which has tracked measures of growth and prosperity over time. The Brookings prosperity index is intended to reflect productivity and the value of labor. Ultimately, increases in productivity and wages are what drive living standards and community well-being. The growth index is based on entrepreneurial activity and employment and includes gross metropolitan product, the number of jobs, and jobs in new firms (less than 5 years old). Employment in new firms is intended as an indicator of investor confidence about a metropolitan area, a proxy for future economic activity. For a third outcome measure, we also use data on the percentage of the population in full-time employment from the US census. In this initial analysis, we examine broadband adoption's relationship to prosperity, growth, and employment. These outcomes are similar to many studied in prior broadband research.

The Brookings Metro Monitor defines economic outcomes as the *change* in prosperity or growth. Prosperity measures are indexed to the year 2000. For all metros, scores are set at 100 in 2000; if scores drop below 100, there has been a decrease compared to 2000. If the index rises above 100, that indicates increased prosperity since the baseline year. For the 50 largest metros and the 16-year time series, the economic prosperity index ranges from a low of 91.36 to a maximum of 139.27 with a mean of 106 and a standard deviation of 6.5. The growth variable is indexed to 2006. Again, all metros begin at 100. When their score falls below 100, growth is decreasing; it is increasing when they rise above 100. Over the 50 largest metros the growth index ranges from a minimum of 62.41 to a maximum of 135.97 with a mean of 88 and a standard deviation of 12. In summary, both our primary explanatory and outcome variables measure change.

To control for the multiple factors that drive the economy, we use multivariate regression with a set of control variables consistent with previous studies on economic outcomes for broadband (Whitacre et al. 2014a; Forman et al. 2012). The time series models control for median income, educational attainment with separate covariates for college graduates, age cohorts and racial and ethnic populations, and core demographic variables in metropolitan areas (US Census Bureau, CPS and ACS). Economic outcomes such as growth and prosperity are strongly influenced by education, especially college-educated workers, and by industry sectors (Moretti 2012; Florida 2017). Education is a factor in rising inequality across communities (Moretti 2012, 99). We also measure the percentage of employed adults in the 14 standard industry categories (NAICS) that the US Census Bureau collects data on (construction industry excluded for reference). As part of this variable we measure the percentage of the population employed in IT jobs.

To control for the rapid expansion of IT over the two decades, we employ statistical models that correct for correlated errors, discussed later. As an added check we also include a series on duration terms that measure time (coded 1 for the first time period in the dataset and 16 for the last time period), as well as time squared and time cubed to measure time as a cubic polynomial.[1] The results are reported with and without the added time duration terms.

Results: Time Series

Table 3.1 presents results from Prais-Winston estimation, a modeling technique intended to handle concerns about autocorrelation of an AR(1) type. By using this estimation technique, we assume the outcome variables of interest—economic growth, prosperity, and full-time employment—are linearly dependent upon previous values of those variables, in addition to a

Table 3.1 Summary (Statistically Significant Models Noted with a Checkmark)

Model	Prosperity index	Prosperity w Durations	Prosperity Robustness	Growth	Growth with Durations	Growth Robustness	Full-Time Employ	Employ with Durations	Employ Robustness
Broadband	✓	✓	✓	✓			✓	✓	✓

stochastic term. This method transforms the dataset to correct for serial cor-relation. Panel-corrected standard errors, a common practice for time series analyses, can appropriately handle heterogeneity between panels, as well as contemporaneous correlation, but serial correlation must be addressed in a different manner (Beck and Katz 1996). This methodological choice prima-rily centers around this concern, in addition to the ability to utilize as much data as possible, given the limited nature of observations between the years 2000 and 2017. Because the data is measured in panels, we also cluster the standard errors by metro area. As a robustness check Table 3.1 also reports the results of models estimated using pooled cross-sectional time series as a comparison (Beck and Katz 1996). The results are comparable, providing added confidence.

Nine models in Table 3.1 and the summary table at the bottom of Table 3.1 illustrate the relationship between broadband subscriptions and prosperity in the top 50 most populous cities/metros in the United States. Again, for all 50 metropolitan statistical areas (MSAs) included in this dataset, the pros-perity index (columns 1–3) begins in the year 2000 with a value of 100 set as a baseline. Each successive year is measured against the economic realities of each metro in the year 2000. Values lower than 100 indicate a shrinking economy, while anything above 100 is indicative of improvement. Column 1 reports the results of the baseline model, while column 2 adds the additional time duration terms. Column 3 provides a robustness check estimating the model as pooled cross-sectional time series. This same modeling pattern is repeated for columns 4–6 when economic growth is the outcome variable and columns 7–9 for full-time employment.

In the first three columns, broadband subscriptions provide a strong, sta-tistically significant predictor of workers' productivity and wages as meas-ured in the Metro Monitor data. Places with a higher percentage of the population connected to the internet, and more change in broadband con-nectivity each year, over time have stronger economic prosperity, controlling for other factors known to be correlated with economic activity. To illustrate the power of this variable, Figure 3.2 graphs the predicted probability of MSA economic prosperity varying broadband subscriptions from less than 5% to 95%, holding all other variables constant at their central tendency. A 10% increase in the change in broadband subscriptions translates to a roughly 2-point gain in prosperity—productivity, wages, and standard of living. The same graphing method is shown in Figure 3.3 based on the model in column

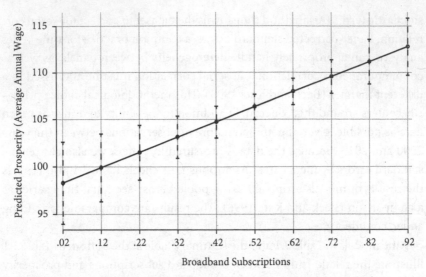

Figure 3.2 Predicted prosperity index by annual change in broadband subscriptions for US metros, Model 1.

Note: Predicted values holding independent variables at central tendencies.

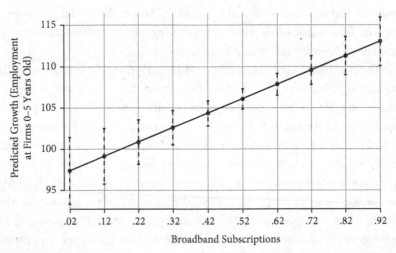

Figure 3.3 Predicted prosperity index by annual change in broadband subscriptions for US metros, Model 3.

Note: Predicted values holding independent variables at central tendencies.

3 (Table 3.1). A 10% increase in change subscriptions translates to a roughly 3-point gain in economic prosperity, all else equal.

The Metro Monitor's index for economic growth includes potential future investment. This variable measures change in a metropolitan area's economy and the level of economic activity. While the change in broadband subscriptions is predictive of increased economic growth when the duration terms are not included (column 4), the result disappears when the time cubic polynomials are included (column 5) and when the alternative time series modeling (pooled cross-sectional time series) is used (column 6). This suggests that broadband subscriptions are not as powerful a predictor for growth compared with prosperity. Nevertheless, the result for the baseline model is significant.

Table 3.1 offers additional insight into the way broadband adoption influences the regional economy by measuring full-time employment. In columns 7 (baseline), 8 (model with duration terms), and 9 (robustness test), the lagged term for broadband subscriptions in the prior year (absolute rate, not change) is a consistent predictor of higher full-time employment in the metro, all else equal. The lagged term measures the percentage of the population with a broadband subscription in the previous year, and this is also statistically significant and substantively important. As shown in Figure 3.4 (left side), using Prais-Winston modeling, a 10% increase in broadband adoption in the prior year translates to a roughly 2-point gain in metro employment. Using pooled cross-sectional time series, a 10% increase in broadband subscriptions equates to, on average, a 3-point increase in full-time employment across metros. These are substantively large findings given the millions of people living in these areas. They show benefits of inclusive broadband use for economic opportunity. Using time series models, the results stand the test of rigorous statistical modeling.

Results: Interaction of Broadband with Young Populations

Models in Table 3.3 offer additional insight into the way broadband access influences regional economies, showing statistically significant interaction effects between broadband adoption (lagged term) and larger millennial populations for predicting prosperity (column 1). Broadband's effect in improving prosperity in America's cities and suburbs is enhanced in regions with a higher share of the population that is aged 25 to 34. The analysis

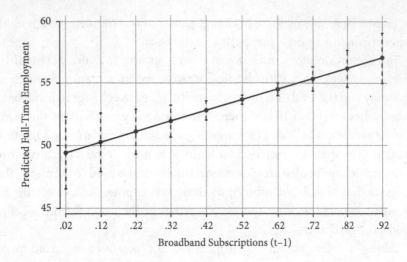

(b)

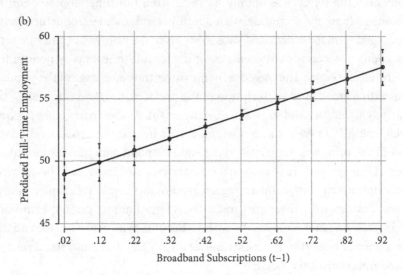

Figure 3.4 Predicted full-time employment by broadband subscriptions for US metros.

Note: Predicted values holding independent variables at central tendencies.

shows that the effect of broadband on regional prosperity is conditioned by the proportion of the population that is millennial—when it is low (20%), broadband still has an effect, but it is slight. For a metro to really unleash its prosperity potential, it needs to have upwards of 30% of its population composed of millennials (Figure 3.5).

Table 3.2 Effect of Broadband Subscriptions on Metro Monitor Prosperity, new firm growth, and full-time employment.

	Prosperity Index			Growth Index			Full-Time Employment		
	(1) Prais-Winston	(2) Prais-Winston w duration	(3) Pooled Cross-Sectional	(4) Prais-Winston	(5) Prais-Winston w Duration	(6) Pooled Cross-Sectional	(7) Prais-Winston	(8) Prais-Winston w Duration	(9) Pooled Cross-Sectional
Broadband Subscriptions (lagged 1 year)	-0.521	5.180	0.124	12.96	7.255	-18.28	10.58***	8.481**	9.499***
	(2.949)	(2.735)	(5.051)	(7.287)	(8.497)	(11.35)	(2.335)	(2.754)	(1.788)
Change Broadband Subscriptions	14.70***	10.95***	17.24***	18.68*	6.249	6.069	-1.261	-3.839	-0.242
	(2.049)	(2.023)	(4.971)	(7.321)	(8.139)	(11.41)	(2.287)	(2.654)	(1.677)
High School Graduates (less high school reference group)	-0.00238	0.129	0.162	-0.182	-0.0293	-0.270	0.172	0.213**	0.0889
	(0.128)	(0.131)	(0.127)	(0.417)	(0.393)	(0.366)	(0.102)	(0.0779)	(0.0561)
Associate's Degree	-0.132	-0.167	-0.439**	-1.900***	-1.448***	-1.151**	-0.0630	-0.170*	-0.157*
	(0.133)	(0.143)	(0.148)	(0.455)	(0.408)	(0.361)	(0.102)	(0.0811)	(0.0610)
College Graduates	-0.0724	-0.0784	-0.233	0.0284	0.203	0.385	0.364**	0.298**	0.397***
	(0.148)	(0.143)	(0.163)	(0.412)	(0.405)	(0.394)	(0.104)	(0.106)	(0.0681)
Professional Degree	-0.334*	-0.184	-0.456	-0.443	0.0495	0.0827	0.181	0.219	-0.404***
	(0.165)	(0.156)	(0.265)	(0.593)	(0.608)	(0.577)	(0.197)	(0.165)	(0.115)
Full-Time Employment	-0.142**	-0.185***	-0.100	0.0633	0.278***	0.884***			
	(0.0443)	(0.0425)	(0.0920)	(0.0901)	(0.0835)	(0.188)			

Continued

Table 3.2 Continued

	Prosperity Index			Growth Index			Full-Time Employment		
	(1) Prais-Winston	(2) Prais-Winston w duration	(3) Pooled Cross-Sectional	(4) Prais-Winston	(5) Prais-Winston w Duration	(6) Pooled Cross-Sectional	(7) Prais-Winston	(8) Prais-Winston w Duration	(9) Pooled Cross-Sectional
Median Household Income	0.000399***	0.000288***	0.000609***	0.000456*	0.000219	0.0000430	0.000217***	0.000126*	0.000330***
	(0.0000652)	(0.0000768)	(0.0000693)	(0.000178)	(0.000194)	(0.000168)	(0.0000475)	(0.0000535)	(0.0000280)
Black (white reference group)	-0.0671*	-0.0486	0.00274	-0.0995	-0.0787	-0.222*	0.0223	0.0149	0.0329
	(0.0257)	(0.0263)	(0.0499)	(0.0901)	(0.0940)	(0.0964)	(0.0250)	(0.0222)	(0.0211)
Asian American	-0.387**	-0.287*	-0.582***	-0.964**	-0.602*	-0.279	-0.437***	-0.356***	-0.362***
	(0.113)	(0.117)	(0.119)	(0.301)	(0.289)	(0.257)	(0.0476)	(0.0479)	(0.0497)
Female (male reference)	0.648	0.589	2.299***	0.743	1.872	4.000*	-0.593	-0.543	-0.366
	(0.612)	(0.595)	(0.687)	(1.699)	(1.622)	(1.804)	(0.624)	(0.459)	(0.304)
Age Cohorts (74 and older reference)									
18-24	0.0210	0.333	-0.438	-0.114	1.952	0.913	0.140	-0.118	0.0219
	(0.406)	(0.394)	(0.355)	(1.646)	(1.424)	(0.931)	(0.253)	(0.234)	(0.159)
25-34	1.129**	1.021**	1.197***	0.550	2.234*	2.071*	0.578*	0.238	0.886***
	(0.328)	(0.356)	(0.295)	(1.279)	(1.007)	(0.838)	(0.241)	(0.247)	(0.130)
35-44	0.569	0.244	0.116	-0.976	-0.601	-1.246	1.040***	0.678**	0.363*
	(0.439)	(0.384)	(0.360)	(1.418)	(1.194)	(1.066)	(0.278)	(0.250)	(0.162)
45-54	0.761	0.731	0.260	-0.111	-1.019	-0.541	-0.416	0.140	0.0908
	(0.525)	(0.525)	(0.387)	(1.349)	(1.184)	(1.028)	(0.314)	(0.275)	(0.173)
55-64	0.799	0.162	-0.361	-5.366**	-3.440*	-2.700*	0.940**	0.392	1.144***
	(0.670)	(0.744)	(0.476)	(1.735)	(1.548)	(1.144)	(0.346)	(0.302)	(0.204)
65-74	2.119***	1.289**	2.734***	3.532**	5.165***	2.510	0.472	-0.726	0.535**
	(0.409)	(0.457)	(0.545)	(1.077)	(1.192)	(1.339)	(0.410)	(0.369)	(0.206)

Industry Sectors (*construction reference*)

Agriculture	1.342	2.044**	3.308***	3.700*	5.129**	1.858	0.830**	1.055***	1.488***
	(0.741)	(0.750)	(0.879)	(1.473)	(1.473)	(1.639)	(0.258)	(0.242)	(0.373)
Manufacturing	-0.247	-0.0455	-0.815*	-0.105	0.214	-1.595*	0.335**	0.480***	0.142
	(0.225)	(0.197)	(0.339)	(0.666)	(0.587)	(0.666)	(0.115)	(0.105)	(0.147)
Wholesale	-0.312	0.100	-1.890*	-0.241	0.447	-1.790	0.141	0.519	-1.691***
	(0.430)	(0.377)	(0.891)	(1.116)	(1.093)	(1.902)	(0.329)	(0.296)	(0.390)
Retail	-0.483	-0.401	-1.453**	-0.711	-0.398	-1.574	-0.0114	0.206	0.592*
	(0.249)	(0.223)	(0.555)	(0.865)	(0.770)	(1.101)	(0.207)	(0.179)	(0.248)
Transportation	0.0738	0.0602	-0.610	-0.173	-0.329	-1.964	0.336	0.244	0.668**
	(0.316)	(0.293)	(0.562)	(0.937)	(0.800)	(1.016)	(0.171)	(0.192)	(0.239)
Information Technology	-0.105	0.398	1.991**	-1.915	-1.336	-4.517***	0.265	0.686*	0.425
	(0.509)	(0.494)	(0.608)	(1.225)	(1.231)	(1.318)	(0.338)	(0.281)	(0.260)
Finance	-0.327	-0.0857	-1.431**	0.345	0.302	-2.064*	0.149	0.470**	0.101
	(0.347)	(0.308)	(0.453)	(0.881)	(0.808)	(0.873)	(0.152)	(0.142)	(0.193)
Professional Services	-0.0236	0.175	-1.282**	-0.0878	0.433	-0.737	0.0828	0.234	0.128
	(0.225)	(0.185)	(0.440)	(0.781)	(0.689)	(0.865)	(0.158)	(0.142)	(0.192)
Education	-0.561	-0.311	-1.252***	-0.356	0.00478	-1.143	0.0502	0.186	0.327*
	(0.301)	(0.281)	(0.371)	(0.644)	(0.626)	(0.729)	(0.151)	(0.142)	(0.159)
Arts	-0.167	0.0927	-0.841*	-0.346	0.253	-1.490*	0.354**	0.528***	0.234
	(0.295)	(0.248)	(0.408)	(0.627)	(0.553)	(0.746)	(0.123)	(0.111)	(0.173)
Public Administration	-0.450	-0.0734	-0.364	0.350	-0.593	-2.973**	-0.0962	0.390*	-0.261
	(0.314)	(0.263)	(0.490)	(0.912)	(0.926)	(0.978)	(0.180)	(0.165)	(0.209)

(*continued*)

Table 3.2 Continued

	Prosperity Index			Growth Index			Full-Time Employment		
	(1) Prais-Winston	(2) Prais-Winston w duration	(3) Pooled Cross-Sectional	(4) Prais-Winston	(5) Prais-Winston w Duration	(6) Pooled Cross-Sectional	(7) Prais-Winston	(8) Prais-Winston w Duration	(9) Pooled Cross-Sectional
Government Services	−0.386*	−0.253	−0.528***	−1.439***	0.0123	−0.210	0.427***	0.212**	0.0764*
	(0.154)	(0.147)	(0.0855)	(0.395)	(0.453)	(0.258)	(0.101)	(0.0790)	(0.0374)
Year (2015)	1.376***	1.320***	1.304*	0.559	1.271**	1.972*	−0.317*	−0.356**	0.455*
	(0.137)	(0.123)	(0.527)	(0.479)	(0.436)	(0.993)	(0.119)	(0.128)	(0.232)
Duration Controls									
Time		3.020***	3.758***		5.588**	2.775		1.129***	
		(0.561)	(0.594)		(1.710)	(2.323)		(0.205)	
Time Squared		−0.308***	−0.216***		−1.603***	−1.198***		−0.0103	
		(0.0574)	(0.0563)		(0.196)	(0.282)		(0.0188)	
Time Cubed		0.0103***	0.00369		0.0926***	0.0757***		0.000309	
		(0.00213)	(0.00226)		(0.00935)	(0.0126)		(0.000749)	
Constant	34.68	25.75	12.41	149.0	−25.79	25.92	8.431	7.585	−6.023
	(38.62)	(36.23)	(45.56)	(107.1)	(93.92)	(110.9)	(31.15)	(22.72)	(20.13)
N	800	800	800	542	542	542	800	800	800
R^2	0.835	0.849		0.814	0.855		0.923	0.936	
Rho	0.984	0.984	0.670	0.934	0.914	0.495	0.843	0.782	0.546
F	51.07	94.86		40.26	87.28		156.7	398.6	

Standard errors in parentheses. Outcome variables from the Metro Monitor, Brookings Institution.

$* p < 0.05$; $** p < 0.01$; $*** p < 0.001$.

(a)

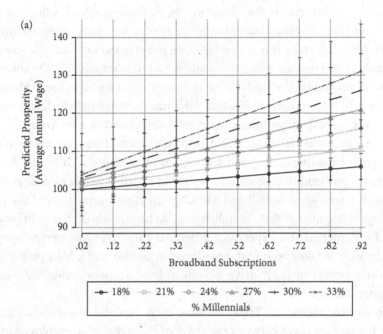

(b)

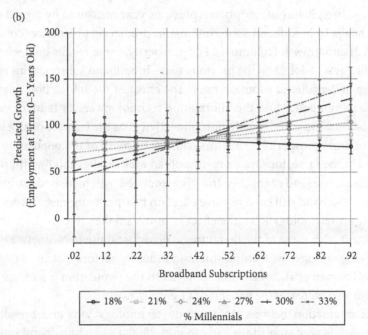

Figure 3.5 Predicted economic outcome by change in broadband subscriptions and millennial populations for US metros.

Note: Predicted values holding independent variables at central tendencies.

Table 3.2 also reports the results for the economic growth values, or the indicator offered by the Metro Monitor as a proxy for entrepreneurial investment. Models 3 and 4 reveal an interesting pattern; the interactions between age cohort and percent with broadband are significant whether the absolute value measured by the lag term or change in broadband is used. Model 3 (Table 3.2), for example, shows that the annual change in broadband adoption on economic growth, measured by employment at new firms, is conditioned by a metro's young population. The interaction term between the change in broadband adoption and age cohorts is statistically significant. Predicted probabilities shown on the right in Figure 3.5 indicate that growth is much lower when broadband subscription is low, regardless of the percentage of population that are millennials. At high levels of change in broadband use, metros still experience growth if they have low percentages of millennials. It takes both high broadband adoption and a high proportion of young people to really drive growth, at least as measured by the Metro Monitor index.

Finally, the models in Table 3.4 show there is also a statistically significant interaction effect between the share of the population employed in IT jobs and broadband subscriptions (previous year measured by lagged term and change in broadband subscriptions) in predicting prosperity (columns 1 and 2) and growth (column 4). For prosperity, these results hold whether one interacts IT jobs and the previous year's broadband subscriptions or the change in broadband adoption rates. The effect of the interaction relationship is more nuanced for the information technology sector than for young populations. Increased broadband connectivity has the largest positive effect on economic prosperity for metros with 3% or more of the population employed in the IT sector. Once a metropolitan area reaches this 3% threshold, broadband use (and change) is strongly associated with increased prosperity. While broadband still has a positive effect on prosperity for metro areas with almost no technology jobs, the effect is much smaller (Figure 3.6). This is consistent with some of the literature on broadband use in businesses—that effects are stronger with employment in industry sectors that use IT intensively (Forman et al. 2012)—but adoption in the population is a measure of use that also increases the benefits of IT employment.

The interaction between information technology jobs and broadband for growth is very large in the early to mid-2000s, when broadband was just starting to transform the economy. Metro areas with 3% or more of the population employed in IT and more broadband subscriptions experienced

Table 3.3 Effect of Broadband Subscriptions on Metro Economic Prosperity and Growth Conditional on Young Populations, 2000–2017

	Prosperity Index		Growth Index	
	(1)	*(2)*	*(3)*	*(4)*
Broadband Subscriptions ($t-1$)	7.555**	−20.37	10.89	−145.7**
	(2.669)	(15.76)	(7.883)	(45.55)
Change Broadband	−28.84	3.795	−176.9*	−0.363
Subscriptions	(15.33)	(2.019)	(67.95)	(7.147)
Interactions				
Millennials ×	1.556*		8.407*	
Change Broadband Subscriptions	(0.726)		(3.220)	
Millennials ×		1.315		7.502***
Broadband ($t-1$)		(0.795)		(2.104)
Demographics				
High School (less high school	0.175	0.172	−0.0502	−0.0406
reference)	(0.131)	(0.133)	(0.380)	(0.387)
Associate's Degree	−0.189	−0.192	−1.476***	−1.473***
	(0.141)	(0.141)	(0.391)	(0.396)
College Graduates	−0.0283	−0.0118	0.211	0.294
	(0.140)	(0.144)	(0.390)	(0.395)
Professional Degree	−0.231	−0.271	0.163	−0.00876
	(0.186)	(0.190)	(0.605)	(0.695)
Full-time Employment	−0.199***	−0.199***	0.249**	0.242*
	(0.0483)	(0.0492)	(0.0858)	(0.0923)
Median Household Income	0.000288**	0.000296***	0.000129	0.000142
	(0.0000862)	(0.0000840)	(0.000184)	(0.000199)
Black (white reference)	−0.0513	−0.0530	−0.0632	−0.0680
	(0.0298)	(0.0301)	(0.0948)	(0.100)
Asian	−0.259*	−0.260*	−0.539	−0.519
	(0.125)	(0.126)	(0.270)	(0.273)
Female	0.705	0.748	1.874	2.011
	(0.683)	(0.691)	(1.552)	(1.650)
Age Cohorts (74 and older reference category)				
Millennials (18–34)	0.232	0.391	−3.366	−2.289
	(0.423)	(0.409)	(2.675)	(1.622)
35–44	0.709	0.684	−0.236	−0.157

Continued

Table 3.3 *Continued*

	Prosperity Index		Growth Index	
	(1)	*(2)*	*(3)*	*(4)*
	(0.435)	(0.418)	(1.104)	(1.095)
45–54	1.016	1.002	−0.608	−0.387
	(0.555)	(0.554)	(1.252)	(1.272)
55–64	0.890	0.872	−2.332	−2.085
	(0.803)	(0.757)	(1.415)	(1.372)
65–74	1.078*	1.108*	4.109**	3.864**
	(0.469)	(0.455)	(1.316)	(1.293)
Industry (construction reference category)				
Agriculture	2.625**	2.609**	5.496***	5.287***
	(0.758)	(0.754)	(1.317)	(1.374)
Manufacturing	0.207	0.178	0.310	0.180
	(0.204)	(0.212)	(0.560)	(0.559)
Wholesale	0.547	0.598	0.538	0.671
	(0.485)	(0.490)	(1.147)	(1.130)
Retail	−0.450	−0.449	−0.320	−0.353
	(0.259)	(0.268)	(0.799)	(0.798)
Transportation	0.225	0.197	−0.169	−0.337
	(0.358)	(0.369)	(0.740)	(0.767)
Information Technology	0.335	0.313	−1.465	−1.545
	(0.536)	(0.546)	(1.223)	(1.198)
Finance	0.234	0.162	0.639	0.381
	(0.324)	(0.339)	(0.705)	(0.740)
Professional Services	0.381	0.354	0.555	0.463
	(0.216)	(0.223)	(0.663)	(0.666)
Education	−0.376	−0.376	0.0226	0.0308
	(0.311)	(0.322)	(0.614)	(0.613)
Arts	0.175	0.160	0.347	0.269
	(0.283)	(0.292)	(0.540)	(0.535)
Public Administration	0.0703	0.0539	−0.416	−0.517
	(0.392)	(0.396)	(0.889)	(0.886)
Government Services	−0.363*	−0.354*	−0.140	−0.128
	(0.166)	(0.167)	(0.397)	(0.396)
Duration Controls				
Time	3.025***	3.151***	4.430*	4.024*
	(0.511)	(0.499)	(1.748)	(1.775)
Time Squared	−0.248***	−0.257***	−1.381***	−1.332***
	(0.0541)	(0.0553)	(0.202)	(0.209)

Table 3.3 *Continued*

	Prosperity Index		Growth Index	
	(1)	*(2)*	*(3)*	*(4)*
Time Cubed	0.00788***	0.00807***	0.0829***	0.0813***
	(0.00208)	(0.00214)	(0.00904)	(0.00917)
Constant	4.777	0.774	74.78	
	(43.31)	(42.98)	(104.7)	(88.82)
N	800	800	542	542
R^2	0.863	0.866	0.871	0.873
Rho	0.980	0.980	0.894	0.891
F	68.80	75.76	70.74	57.52

Standard errors in parentheses. Outcome variables from the Metro Monitor, Brookings Institution.
* $p < 0.05$; ** $p < 0.01$; *** $p < 0.001$.

steady and significant growth. Moving from 12% of a metro's population on-line to 32% shifts this entrepreneurial growth by 5 points on the index for metros with 6% of the population working in IT (see Figure 3.6, right-hand side). The same 20-point increase in the population online for non-tech cities resulted in just a 1-point change. As the economy has transformed in the digital era with 83% of the population online today, the differences in growth between tech hub cities and nontech hubs seem to have leveled off. Research on business use of broadband covered an earlier era in the diffusion of broadband, so these decreasing effects for IT employment over time are useful for understanding the changing context when the impacts of technology may be broader.

Conclusion: Digital Human Capital for Economic Opportunity

Metropolitan areas represent regional labor markets and hubs for innovation in the economy. While some earlier studies have examined the effects of broadband deployment or internet use by firms in metropolitan areas, no previous research has definitively modeled the impact of broadband use by the population on economic outcomes over time. Measures of growth, prosperity, and full-time employment addressed here shape the regional environment for economic opportunity.

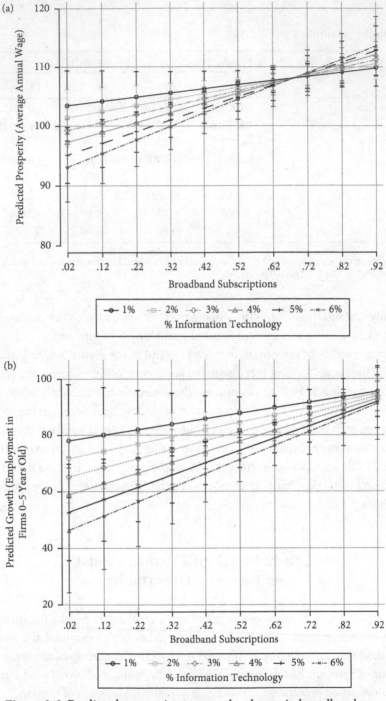

Figure 3.6 Predicted economic outcomes by change in broadband subscriptions and information technology jobs for US metros.
Note: Predicted values holding independent variables at central tendencies.

Using new data, we examine broadband adoption and economic change over nearly two decades in the 50 largest US metros. Time series analysis demonstrates that broadband adoption in the population is causally related to metropolitan economic outcomes. Broadband adoption does predict economic prosperity, growth, and employment in metros, though in somewhat different ways. Broadband is a strong predictor of prosperity (productivity, wages, and standard of living) and full-time employment. We uncover these results using rigorous time series models across the 50 largest metros, measuring both the absolute levels of broadband adoption and change in broadband adoption from the previous year simultaneously.

A broader picture of broadband adoption emerges when its impact is considered in combination with other trends. Broadband subscriptions in the population are significant on their own, but also interact with other factors, such as IT employment and the region's share of millennials. The metropolitan areas where high-speed internet is more widely used, where IT is a significantly larger proportion of the labor force, and where millennials have a larger footprint are also the metropolitan areas where prosperity and entrepreneurial activity are at the highest levels.

Broadband subscriptions increase prosperity more in communities with significant younger populations, at least 30% aged 25 to 34. These relationships reflect recent trends in the clustering of industries and populations in the United States; for example, finalists for the new Amazon headquarters tended to be cities with concentrations of educated millennials (Frey 2018). High levels of broadband adoption increase growth even when the percentage of millennials is low, but the combination raises the bar significantly. Interactions between broadband subscriptions and IT employment show that changes in prosperity are modest until there is at least 3% of the population in IT occupations. Once a metropolitan area reaches this 3% threshold, broadband use (and change in use) is strongly associated with increased prosperity. While broadband still has a positive effect on prosperity for metro areas with almost no technology jobs, the effect is much smaller.

The interactive effects of broadband adoption and IT employment for growth differ over time. In the early to mid-2000s, moving from 12% of a metro's population online to 32% shifts this entrepreneurial growth by 5 points on the index for metros with 6% of the population working in IT. The same 20-point increase in the population online for nontech cities resulted in just a 1-point change. In the latter period, however, the effects of the concentration of IT employment have leveled off.

Table 3.4 Effect of Broadband Subscriptions on Metro Economic Prosperity and Growth Conditional on Population Employed in Information Technology Jobs, 2000–2017

	Prosperity Index		Growth Index	
	(1)	(2)	(3)	(4)
Broadband Subscriptions	−0.886	−8.209	11.06	−13.45
(t − 1)	(2.916)	(4.146)	(7.399)	(11.20)
Change Broadband	4.832	12.91***	2.290	15.21
Subscriptions	(3.682)	(2.019)	(17.51)	(7.693)
Interactions				
Information tech ×	3.140**		6.062	
Change Broadband	(1.110)		(5.486)	
Information Tech ×		2.824*		9.486**
Broadband Subscription (t − 1)		(1.302)		(3.465)
Demographics				
High School	0.0880	0.0840	−0.155	−0.0990
	(0.137)	(0.143)	(0.419)	(0.427)
Associate's Degree	−0.0659	−0.0685	−1.863***	−1.810***
	(0.137)	(0.138)	(0.460)	(0.459)
College Graduates	−0.00610	−0.0122	0.0746	0.119
	(0.156)	(0.153)	(0.408)	(0.402)
Professional Degree	−0.293	−0.304	−0.402	−0.376
	(0.168)	(0.166)	(0.600)	(0.594)
Full-Time Employment	−0.132**	−0.130**	0.0745	0.0902
	(0.0456)	(0.0462)	(0.0907)	(0.0884)
Median Household Income	0.000379***	0.000385***	0.000433*	0.000430*
	(0.0000642)	(0.0000634)	(0.000177)	(0.000177)
Black	−0.0637*	−0.0651*	−0.102	−0.105
	(0.0268)	(0.0268)	(0.0882)	(0.0856)
Asian	−0.354**	−0.356**	−0.946**	−0.909**
	(0.123)	(0.123)	(0.295)	(0.290)
Female	0.490	0.589	0.703	0.950
	(0.648)	(0.646)	(1.688)	(1.725)
Age Cohorts				
18–24	0.00716	0.0354	0.00936	0.190
	(0.457)	(0.450)	(1.651)	(1.654)
25–34	1.354**	1.348**	0.806	1.091
	(0.393)	(0.396)	(1.265)	(1.312)

Table 3.4 *Continued*

	Prosperity Index		Growth Index	
	(1)	(2)	(3)	(4)
35–44	0.647	0.640	−0.867	−0.850
	(0.492)	(0.485)	(1.418)	(1.462)
45–54	0.873	0.881	−0.0752	−0.0200
	(0.575)	(0.573)	(1.358)	(1.391)
55–64	0.904	0.916	−5.260**	−5.154**
	(0.732)	(0.706)	(1.719)	(1.725)
65–74	2.278***	2.241***	3.640**	3.615**
	(0.418)	(0.418)	(1.061)	(1.125)
Industry				
Information Technology	−2.147*	−1.911*	−6.454	−8.745**
	(0.869)	(0.859)	(3.536)	(2.539)
Agriculture	1.293	1.302	3.602*	3.483*
	(0.774)	(0.787)	(1.512)	(1.524)
Manufacturing	−0.241	−0.274	−0.119	−0.203
	(0.221)	(0.227)	(0.671)	(0.673)
Wholesale	−0.334	−0.332	−0.300	−0.340
	(0.436)	(0.434)	(1.140)	(1.143)
Retail	−0.496	−0.500	−0.705	−0.725
	(0.249)	(0.254)	(0.868)	(0.894)
Transportation	0.107	0.0414	−0.166	−0.345
	(0.333)	(0.332)	(0.941)	(0.935)
Finance	−0.238	−0.265	0.362	0.331
	(0.341)	(0.352)	(0.874)	(0.869)
Professional Services	0.0116	−0.0183	−0.0924	−0.171
	(0.231)	(0.233)	(0.784)	(0.811)
Education	−0.554	−0.587	−0.359	−0.459
	(0.313)	(0.308)	(0.640)	(0.646)
Arts	−0.228	−0.236	−0.394	−0.487
	(0.308)	(0.303)	(0.637)	(0.642)
Public Administration	−0.405	−0.459	0.349	0.209
	(0.325)	(0.322)	(0.908)	(0.925)
Government Services	−0.374*	−0.367*	−1.403***	−1.351**
	(0.170)	(0.168)	(0.401)	(0.397)
2015 Year Dummy	1.294***	1.356***	0.495	0.575
	(0.134)	(0.134)	(0.479)	(0.478)

Continued

Table 3.4 *Continued*

	Prosperity Index		Growth Index	
	(1)	(2)	(3)	(4)
Constant	34.79	31.38	154.0	143.6
	(41.68)	(41.47)	(108.8)	(107.1)
N	800	800	542	542
R^2	0.828	0.831	0.821	0.828
Rho	0.985	0.985	0.929	0.924
F	58.39	54.15	37.40	41.46

Standard errors in parentheses.
* $p < 0.05$; ** $p < 0.01$; *** $p < 0.001$.

While broadband adoption matters for growth in some models, the relationship to prosperity is more consistent. The prosperity index is focused on workforce and population outcomes, measured through productivity, wages, and standard of living; the stronger results for prosperity make sense for broadband use as a dimension of human capital in the workforce. A younger population may also represent a form of digital human capital, given the greater propensity of young people to be engaged online. Like broadband adoption, age may signal capacities in the workforce important beyond a single sector.

Many of the positive results for earlier studies of broadband deployment or business use were replicated with our measures of broadband adoption over time (Lehr et al. 2006; Kolko 2010; Whitacre et al. 2014a; Forman et al. 2012). Our findings are not necessarily in conflict with the story of rising inequality across metros and the prominence of the superstar cities. Employment in IT, which has concentrated in a relatively few regions, does indeed interact with broadband adoption to influence economic outcomes, though the effects are stronger earlier in broadband's diffusion. But the agglomeration of the technology industry is not the whole story.

Digital human capital matters too and can provide resources for inclusive growth and resilience as economic change continues (West 2018), maximizing the talent in regional labor markets (Parilla 2017). Education and job training are important dimensions of human capital, but the findings on broadband adoption, for metros as well as counties, suggest that the ability to participate in society online is also an aspect of human capital for the future.

Access to technology and digital literacy share some striking parallels with access to public education and literacy in the 19th and 20th centuries, both for building human capital and for promoting greater equality of opportunity (Goldin 1998).

The next two chapters examine patterns and impacts for neighborhoods and states. Digital human capital matters more when it is widespread throughout communities, and in the next chapter we examine spatial disparities at the neighborhood level and policies for inclusive broadband use. Beyond these local communities, there is evidence that the information networks supported by digital human capital facilitate policy innovation at the state level, which in turn affects the context for local economic development and broadband policy.

4

Smart Cities and Neighborhoods

With Bomi Lee

While metros form the nation's urban economies, examining broadband use at the metropolitan level is insufficient to answer pressing policy questions for local innovation and place-based inequality. The metropolitan scale obscures the deep disparities and need that exist within highly connected centers of the digital economy like San Jose as well as cities like Detroit where disadvantage is prevalent citywide. Many neighborhoods within major cities have rates of broadband adoption that lag far behind the nation and surrounding metros. Urban zip codes range from those where broadband adoption is nearly universal to those where only one in four residents has any type of broadband subscription. We track the digital footprint of segregation and concentrated poverty in the nation's largest cities, where many rely on cell phones as their primary way to go online.

Addressing policies for broadband and urban innovation also requires attention to cities and their local governments, as most metropolitan regions have little power or policy authority on their own.[1] One manifestation of local innovation is the global smart cities movement, which has inspired city governments, both large and small, to imagine new uses for information technology (IT) embedded in "infrastructures, architecture and everyday objects" (Townsend 2013, 15). Featuring sensors, gigabit broadband, artificial intelligence, and data, smart cities have a variety of aims and a multitude of definitions (e.g., Albino et al. 2015; Nam and Pardo 2011). The "smart" label captures notions of learning and innovation, experimentation, future-thinking, entrepreneurialism, and creativity through broadband-enabled applications (Albino et al. 2015; Nam and Pardo 2011). Boorsma (2018) cites the goals promoted by early advocates:

> A smart community is a community with a vision of the future that involves the application of information and communication technologies and broadband infrastructures in a new and innovative way to empower

Choosing the Future. Karen Mossberger, Caroline J. Tolbert, and Scott J. LaCombe, Oxford University Press. © Oxford University Press 2021. DOI: 10.1093/oso/9780197585757.003.0004

its residents, institutions and regions as a whole. Importantly, a smart
community is not primarily focused on technology. . . . Companies and
governments that take advantage of these new technologies will create jobs
and economic growth as well as improve the overall quality of life within
the communities in which they take part. (Boorsma 2018, 141–142)

Empowering residents, institutions, and regions relies not just on cutting-
edge technologies but, as Nam and Pardo (2011) explicitly argue, on the
social and human capital in cities, including digital skills for widespread
participation.

This chapter examines patterns of broadband adoption within the 50
most populous cities, many of which have been at the center of the smart
city movement. Economist Ed Glaeser (2011) contends that large cities are
the most innovative, and so their collective human capacity for innovation
is worth examining more closely here. Big cities have the density and diversity within them to support complex economies and other resources for
innovation, such as institutions of learning, discovery, and culture. Their
neighborhoods are racially and ethnically diverse, but also varied in their
access to opportunity. Chetty and colleagues (2018) discover that economic
mobility for the 20 million children they studied differed markedly across
a few miles or even a few blocks within the same city (Chetty et al. 2018).
Given findings in the prior two chapters, understanding the context of urban
broadband use across neighborhoods and demographic groups is necessary
to promote widespread and inclusive economic innovation within cities and
metros.

The next section begins with a discussion of differences across and within
the 50 largest cities, using data from the 2017 American Community Survey
(ACS). The 2017 ACS marks the first time that the US Census Bureau has
released data on internet use at the neighborhood level. While this is available only for a single point in time, this data allows us to describe place-based
inequalities within cities, as well as to draw cross-city comparisons. We
map this data at the zip code level to show how the growing phenomenon
of smartphone-dependent internet use is structured by geography, in cities
with both high and low broadband adoption.

As Nam and Pardo (2011) indicate, smart cities also must develop innovative policy and governance. Examples of major cities that have devoted
resources to digital inclusion as part of the local technology agenda are
discussed. We describe evaluation research for one effort, Chicago's Smart

Communities, which demonstrates that neighborhood-level change is indeed possible through such programs. Local governments can play an important role in promoting broadband adoption for inclusive innovation and growth.

We begin by comparing digital inequality in two cities at opposite sides of the spectrum: Detroit, which ranks lowest among the 50 largest cities for fixed broadband adoption, and San Jose, which is in the top 3 cities. With very different citywide numbers, both cities are striving to build their digital human capital. Detroit has appointed a digital inclusion director to lead local initiatives, and San Jose has passed a local plan and leveraged funding to address gaps in its community.

Cities and Their Neighborhoods: Detroit and San Jose

Chapter 1 discussed broadband adoption in two midsized cities: Flint, Michigan, and Sunnyvale, California. Comparison of the nation's 50 largest cities reveals inequalities across and within cities, patterned by geography as well as demographics. While lack of high-speed infrastructure is one factor influencing low rates of broadband adoption in some rural counties, broadband internet service is generally available in all the nation's largest cities (Tomer et al. 2017). Urban disparities reflect poverty and other social inequalities, and a lack of access to digital information perpetuates these other forms of disadvantage.

Detroit's Nascent Recovery

Like Flint, its neighbor to the north, Detroit became an emblem of urban misfortune. After decades of decline, a precipitous crash in the 2008 recession, and the largest municipal bankruptcy in the nation, Detroit has made progress in its recovery. IT is integral to the city's revitalization, including investments by Quicken Loans and Google. The regional chamber of commerce reports that since 2009, the growth of the IT industry in Detroit has outpaced the state of Michigan and the nation (Detroit Regional Chamber n.d.). Significantly, the chamber still advertises Detroit as the auto capital of the world (Detroit Regional Chamber n.d.); increasingly, however, those products are bundles of artificial intelligence on wheels. The lines between the old manufacturing economy and the new information economy are increasingly blurred.

Following a period when the city was unable to keep the streetlights on, the landscape looks more hopeful. Detroit's downtown and some nearby areas are experiencing a fresh infusion of activity. A new arena and surrounding entertainment district are served by a light rail that runs along the main thoroughfare of Woodward Avenue between downtown and midtown. Restaurants and up-market condos flourish downtown and along the riverfront. Historic Corktown was already booming before Ford Motor Company purchased the abandoned Michigan Central railway station there for mixed-use redevelopment (Saunders 2018; Applebome 2016; Boudette 2018). New projects are thriving in downtown and other central areas.

Detroit's rebound in a few short years since the recession has been remarkable, but it has not yet reached many of the neighborhoods beyond the core (Applebome 2016; Reese and Sands 2017). The city of Detroit had the lowest rate of wireline home broadband subscriptions among the 50 largest cities in 2017, according to the ACS. Only 48.3% of the population had broadband connections other than cell phones, compared to 70.6% for the nation (2017 ACS, five-year estimates)[2]—around three-quarters of the national average. If cell phones are included, then Detroit's rate of internet connectivity was at 67.5% in 2017, compared to the national average of 78.1%. Thus, over half the population lacked a fixed home broadband connection, and one-third had no personal internet access at all, even on a smartphone.

Detroit also had the lowest median household income of the 50 largest cities, at only $27,838 in 2017, compared with the US median of $60,336 and $58,411 for the Detroit metro area. The percentage of the population that was employed in 2017 was second lowest at 59.3% (Tucson, which was at 58% ranked last). Low broadband connectivity is one indicator of broader distress, and the lack of connectivity could also hold back the city's recovery.

Yet Detroit was a cradle of innovation in the early 20th century. The automobile industry did more than develop a faster mode of transportation. It introduced the assembly line and boosted wages to $5 a day. The city gave rise to a wave of industrial unionism that increased pay and swelled the ranks of the middle class. The automobile as a technology spurred policy innovations and social transformation on a national scale, including the interstate highway system and suburban expansion. Like Silicon Valley, Detroit influenced trends in society, beyond its renown for producing a particular commodity.

Detroit's path of development, however, had social and economic consequences, and so will its current recovery. Mass production relied on less skilled labor, creating a deficit of human capital over time (Glaeser 2011, 42–43). Just as Detroit lags in income, in comparison to the rest of the 50 largest cities, it has the lowest percentage of the population with a college degree—just 14.6%, compared to first-place Seattle at 62.6%. This is an incredible 48-percentage-point difference. Suburban sprawl in Detroit and other US cities, made possible by the automobile, encouraged racial and economic segregation. Metropolitan fragmentation, the division of metropolitan areas into sometimes hundreds of independent jurisdictions, meant that businesses and more affluent residents could easily move to more exclusive communities for lower tax rates and superior services, avoiding the cost of supporting services for the poor (Savitch and Adhikari 2017). This also facilitated racial segregation between cities and suburbs in metropolitan areas, a trend highly visible in the Detroit region. The greater mobility fostered by the age of the automobile eventually led to the demise of Detroit's population numbers as well. Between 1950 and 2008, Detroit lost over a million residents, sinking to less than half its former size (Glaeser 2011, 41).

Detroit has survived and embarked on a new beginning. It is one that is marked by uncertainty, however, about the inclusiveness of the recovery and the future. Scholars have argued that Detroit's high-tech revival may be widening inequality for residents who remain without jobs and who lack the appropriate skills for the knowledge economy (Reese and Sands 2017). Digital human capital is clearly one issue for the path forward.

Digital Inclusion in San Jose

Far from Michigan's auto corridor, Silicon Valley embodies the concentration of innovation in the information and data-driven economy. San Jose is the largest city in the region, and though considered a more affordable alternative than its neighbors (Urban et al. 2018), its overall wealth and rates of broadband adoption citywide far outstrip Detroit's. Median household income, at $45,669, was significantly higher in 2017 than in the Motor City (median income in San Jose metro was $117,000). San Jose's percentage of college graduates was triple Detroit's, at 43%. In San Jose 83.9% of the population had broadband at home in 2017, and 91.4% of the city's residents had some connectivity, with cell phones included. This is more than a

35-percentage-point difference from Detroit for fixed broadband, and a 24-percentage-point difference if cell phones are counted.

Yet, the city of San Jose launched a digital inclusion initiative in 2018, pointing out that it has one of the highest rates of income inequality in the nation and an estimated 95,000 residents who lack broadband at home (Douglas 2018; San Jose n.d.). Over 40% of households with annual incomes under $20,000 are without home broadband connections (San Jose Digital Inclusion Fund n.d.; Speed Up San Jose n.d.). The city's goal is to provide free or low-cost internet at speeds of one gigabit or more in low-income communities, along with training and affordable computers.

In 2018, the city reached an agreement with 5G wireless providers that included contributions for the Digital Inclusion Fund and in-kind public benefits such as smart lighting and community Wi-Fi. The city negotiated funding and other benefits in return for use of public utility poles for building new, fast 5G networks in the largest small cell deployment in any US city (Douglas 2018). Over a 10-year period, the companies will contribute $24 million to the fund, which will be used to connect 50,000 households with affordable plans and to offer library programs. According to the mayor's Smart City Vision, San Jose is striving to "ensure all residents, businesses, and organizations can participate in and benefit from the prosperity and culture of innovation in Silicon Valley" (San Jose n.d.).

Actions taken by the Federal Communications Commission (FCC) will prevent other cities from pursuing similar funding schemes. In September, 2018, the FCC issued an order to limit the fees that local governments can charge for public 5G small cell deployment to "no greater than a reasonable approximation of their costs for processing applications and for managing deployments in the rights-of-way" (FCC 2018b).[3] Additionally, the order imposed limits on the time that local governments can take to respond to company requests, making permits automatic if cities don't meet the 60- or 90-day deadlines. The order effectively blocks agreements like San Jose's that leverage funds for digital inclusion programs. It hampers the ability of municipalities to review and regulate deployment. In response, Portland and San Francisco filed a lawsuit in the Ninth Circuit Court of Appeals to challenge the FCC's ruling.

As this chapter will show, municipal governments around the country have taken actions to support broadband adoption in their cities, recognizing that inequalities persist even in communities that are highly connected overall. While the city's demands were criticized by some as an impediment to the rollout of 5G in the Bay Area (Layton 2018), San Jose's digital inclusion strategy has been honored with awards by state and local government organizations (Nyczepir 2018).

Comparing Broadband in the 50 Largest Cities

San Jose made its Digital Inclusion Fund a local policy priority, though in the 2017 five-year ACS estimates it ranked #3 among the 50 largest cities for broadband (with or without cell phone use), as shown in Appendix Table 4.1. The 91.4% of residents with any type of broadband subscription was just one percentage point behind Raleigh, North Carolina's 92.4% and one-tenth of a point behind San Diego, California's 91.5%. The top-ranked city for fixed broadband subscriptions (not including cell phones) was Seattle, the home of Microsoft, Amazon, and other tech giants. In 2017, 85.2% of Seattle's population had fixed broadband.

Disparities across the 50 largest cities are striking, as shown in Appendix Table 4.1. Top-ranked Seattle and #50 Detroit had a 38-percentage-point gap in fixed broadband subscriptions. If internet access through cell phones is included in the definition of broadband (as we have done in other chapters of this book), differences across cities narrow somewhat. Still, there is a 26-percentage-point difference between Raleigh, North Carolina, at 92.4% with broadband (wired and cell phones) and Miami at 65.9%. Considering that 34% of Miami's households lack any kind of home or mobile access to the internet, the numbers are sobering.

While such gaps across urban areas may be unexpected in a digitally enabled society, the relative ranking of cities is unsurprising. Most top 10 cities for fixed broadband adoption are established or rising tech hubs: Seattle, San Diego, San Jose, Raleigh, Virginia Beach, Colorado Springs, San Francisco, Portland, Denver, and Austin. Places lingering at the bottom of the list tend to be in the Midwest or South, and all the bottom 10 are majority-minority cities: Detroit, Miami, Memphis, Milwaukee, New Orleans, Philadelphia, Baltimore, El Paso, Dallas, and Houston (2017 ACS, five-year estimates).

When access to the internet through cell phones is included in the definition of broadband, city rankings shift slightly. The 10 lowest-ranked cities for fixed broadband subscriptions all have double-digit percentages of residents with only smartphone access. This compares with the ACS 2017 national average of 7.5%. Seven of the 10 lowest-ranked cities for fixed broadband connections have over double this rate—more than 15% of residents who depend on smartphones for personal internet access. Detroit, the lowest-ranked city for fixed broadband, tops the list for cell-only internet use at 19.2%. Even with the inclusion of smartphone use, the seven lowest-ranked cities remain at the bottom, with from a quarter to a third of their populations with no internet subscription at all.

Only 4 of the 50 largest cities have rates of cell-only internet use at or below the national average. Clearly this is an important form of access today in low-income communities. Smartphones have in some ways narrowed gaps in internet use, as their adoption is similar across racial and ethnic groups. Young, less educated, low-income, African American, and Latino internet users are more likely to rely on cell phones for their primary form of internet access (Pew Research Center 2019).

City Broadband Subscriptions for Racial and Ethnic Groups

Data for the 50 largest cities in the 2017 ACS demonstrates that disparities remain for urban African Americans and Latinos, even when cell phone use is included in measures of broadband adoption, but the gaps have narrowed from a decade earlier (Appendix Table 4.2).

There is not a single one of these large cities where African American or Latino broadband adoption equals that of Asian Americans or non-Hispanic whites, even with cell phone use included (Virginia Beach is close for Latinos, but has a nearly 10-percentage-point gap between African Americans and non-Hispanic whites). Still, these groups have higher rates of internet use in cities where there is a high proportion of the population online overall. In Virginia Beach, 90.2% of Latino households have some type of broadband subscription, compared to only 57.7% in Detroit. In San Jose, 87.4% of African American households have either home broadband or smartphones, in contrast to only 51.9% in Miami. Tech hub cities may have environments that encourage and support technology use, so that all groups fare better as overall connectivity in the city rises. Just as there are economic spillovers and multipliers from the innovation ecosystems in such cities (Moretti 2012), there may be informal learning and positive spillovers in technology use.

The Digital Footprint of Segregation and Concentrated Poverty within Cities

Neighborhood-level data is needed to track the effects of economic, racial, and ethnic segregation on broadband use and opportunity within cities. Neighborhoods in general have become more segregated by income over the past 40 years if not by race (Owens 2016; Bischoff and Reardon 2014). Within metropolitan areas, approximately 25% of the population of central cities lives in areas of concentrated poverty; despite the growing suburbanization of poverty, this compares with only 7% of suburban populations (Kneebone and Holmes 2016). African Americans and Latinos are still more likely to live in areas of spatially concentrated poverty than low-income non-Hispanic whites (Kneebone and Holmes 2016; Logan and Stults 2011; Lewis and Hamilton 2011; Trounstine 2018).

As Kneebone and Holmes (2016) indicate, those who are poor and live in high-poverty areas experience a double burden. In addition to having limited incomes, they must contend with environments that often have fewer job opportunities and less upward mobility (Kneebone and Holmes 2016). Neighborhood disadvantage for internet use is consistent with research on "neighborhood effects" in other policy areas such as health, education, and employment (Jargowsky 1997; Massey and Denton 1993; Wilson 1987). Previous research has shown that neighborhood context influences home broadband adoption. Community income matters for home access and use across racial and ethnic groups, controlling for both neighborhood and individual characteristics. In fact, it is neighborhood poverty that accounts for the gaps in home broadband adoption between African Americans and non-Hispanic whites. Black residents in more affluent neighborhoods do not differ from similarly situated whites in home internet adoption; it is those who live in poor neighborhoods who have more limited digital opportunities (Mossberger et al. 2006). Residence in poor, predominantly African American and Latino areas magnifies barriers to technology use, controlling for other factors (Mossberger, Tolbert, Bowen, and Jimenez 2012).

Neighborhood environment may matter for many reasons. An "ecology of support" (Rhinesmith 2016) in localized social networks and neighborhood institutions is important for broadband adoption and skills. This includes resources in libraries and other community centers (Rhinesmith 2016; Gangadharan and Byrum 2012). Institutional racism, including disparities in employment and education in poor communities of color, may also affect internet skills and opportunities for learning. Unequal education has long-term effects, as residents of racially segregated low-income neighborhoods are likely to have grown up in similar areas, with underresourced schools (Holloway and Mulherin 2004). Residents in poor communities of color may have fewer opportunities to find jobs where they can develop skills using technology (Kaplan and Mossberger 2012). Goods and services sold in poor neighborhoods are often more expensive, credit is less available, and consumer information about some practices leading to unanticipated costs may be less prevalent, making it more difficult to afford broadband (Caplovitz 1967; Brookings Institution and Federal Reserve 2008; Dailey et al. 2010).

One Chicago study on building a "Smart Black Tech Ecosystem" demonstrates how local contexts can encourage or limit opportunity and innovation (Wilson et al. 2019). Chicago is a more diverse city than many of the existing tech hubs, with a growing technology industry presence, many African American professionals, and more women of color who are starting tech businesses. Yet, in examining the prospects for building a Black tech ecosystem in Chicago, the report finds that "sustained patterns of racial segregation—which continue to concentrate the black community on the south and west sides of the city as well as in a few southern suburbs—blunts this advantage" (Wilson et al. 2019, 8). Unequal funding and resources for schools limited the pipeline for technology professions, and only 7 of the 32 technology incubators and accelerators were located within five miles of a predominantly African American neighborhood (Wilson et al. 2019). Encouraging digital human capital at multiple levels—from widespread broadband use to tech start-ups—includes addressing these social costs of racial segregation and concentrated poverty in urban neighborhoods. The larger costs to society are a loss of human capital and innovation that affects the economy as a whole.

Measuring Urban Neighborhood Broadband Use

The neighborhood-level data in the ACS differentiates between fixed and mobile broadband, unlike the ACS data by race and ethnicity in Appendix Table 4.2.[4] There are real differences in the capacity to engage in many internet uses, comparing fixed broadband to internet use primarily on cell phones. Smartphone-dependent internet users enjoy more access to the internet than those who rely on public access, but they are also less connected internet users compared with those who have home broadband or multiple forms of access (Mossberger et al. 2013). With smaller screens and keyboards, cell phones are less useful for reading text-heavy websites, doing schoolwork, filling out forms, or composing documents. According to surveys conducted by the Pew Research Center, smartphone-dependent internet users often have low-cost plans with data caps that limit activities online or result in loss of service (Smith and Page 2015). Such data caps limit use of cell phones as hot spots to connect laptops or tablets.

What is the impact of this trend toward mobile-dependent internet use in poor neighborhoods? The limitations of this mode of access were readily apparent as learning became remote in the pandemic, affecting K-12 students and low-income college students alike. Even prior to this, however, a 2018 citywide survey in Seattle indicated that smartphone-dependent internet users were less likely to feel that they had adequate access (Pacific Market Research 2019). Research on Chicago's neighborhoods demonstrates the dual role of smartphone access. In the city's poor, segregated neighborhoods, smartphone use increased the range of economic, political, and civic activities online for residents (Mossberger et al. 2017)—the types of digital activities that others have called capital enhancing (DiMaggio et al. 2001; Hargittai 2002). Cell phone use for these purposes demonstrates determination to go online even with more limited resources, a finding borne out by other researchers who conducted focus groups with cell-only internet users in Detroit (Fernandez et al. 2020). Yet, it is also true that those with fixed broadband at home engage in an even higher number of these activities online, controlling for other factors (Mossberger et al. 2013, 2017). As one study by the Brookings Institution observed, fixed broadband remains "a critical, in-home gateway to the content, applications, and services that enable households to participate in a digital economy" (Tomer et al. 2017).

Yet reliance on smartphones to go online has increased nationally in recent years (Anderson 2019). In their study of Los Angeles, Galperin and colleagues (2017) conclude that a digital underclass has emerged in that city, patterned by geography as well as demography, and based on internet users who have only smartphones and no home broadband. The neighborhood-level analysis in this chapter demonstrates more general relevance for these observations about the less connected (Mossberger et al. 2013), who are internet users without fixed broadband.

With more granular data than previously available, the 2017 ACS estimates permit new comparisons of broadband use across neighborhoods, differentiating between fixed and wireless connections in households. While neighborhood boundaries differ across cities, or are often informal, researchers have used census tracts or zip codes to represent neighborhoods in urban areas. Census tracts can be very small in densely populated cities, with multiple tracts clustered into neighborhoods. In Chicago, for example, there are 866 census tracts, but 85 zip codes and 77 official community areas (the latter are used for planning). Zip codes are closer to the scale of neighborhoods or districts as defined in many cities and are used here for comparisons within and across cities.[5]

The 2017 ACS data reveal that cell phone–reliant internet use varies by neighborhood within cities. Figures 4.1 through 4.3 show broadband subscriptions of any type (including mobile) on the left and cell-only subscriptions on the right, by zip code, for the three cities with the highest levels of fixed broadband adoption. City-level data suggest that internet use is nearly universal in the three top cities (Seattle, San Diego, and San Jose), but neighborhood data tell a different story. The maps of fixed broadband and cell-only use appear to be mirror opposites. Neighborhoods with the lowest rates of broadband subscriptions have the highest percentage of the population relying on cell phone–only internet. Neighborhoods with the highest rates of broadband subscriptions have the lowest reliance on cell phone–only internet. *This is the digital footprint of concentrated poverty and segregation across urban neighborhoods.*

In Seattle, Figure 4.1 (left) shows broadband subscriptions of any type, with darker shading for zip codes with higher rates of subscriptions. Although the city of Seattle ranks #1 for fixed broadband at 85.2% and 91.3% of residents have some type of broadband subscription, zip codes within the city vary from 66% to 92% for all types of broadband. Figure 4.1 (right) tracks cell phone–only internet subscriptions by zip code. The darker areas, which have higher rates of cell phone–only internet use, are nearly mirror images of the zip codes with higher broadband adoption. Although cell-only internet use is lower in Seattle than other cities, at 6.5%, zip codes in the city have between 3.2% and 11.3% of households that rely on smartphones to go online. Broadband subscriptions overall are higher in north Seattle, and smartphone-only internet use is higher on the city's south side.

In San Diego, fixed broadband is 83.9% (91.5% for all types of broadband). Figure 4.2 (left) shows that broadband subscriptions of any type are lowest in the far northeast and in parts of the south side. San Diego's zip codes range from 61.6% to 93.7% of households with a broadband subscription. Cell-only

(a)   (b)

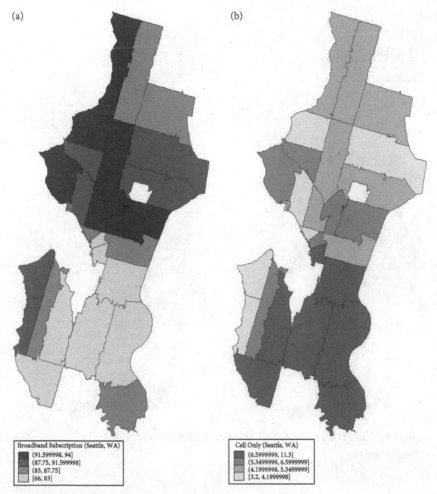

Broadband Subscription (Seattle, WA)
■ (91.599998, 94]
■ (87.75, 91.599998]
□ (83, 87.75]
□ [66, 83]

Cell Only (Seattle, WA)
■ (6.5999999, 11.3]
■ (5.3499999, 6.5999999]
□ (4.1999998, 5.3499999]
□ [3.2, 4.1999998]

Figure 4.1 Seattle, Washington (ranked #1 for fixed broadband subscriptions).

internet subscriptions vary somewhat more than in Seattle, with zip codes reporting between virtually zero and 16.1% cell-only use. Like Seattle and the other metros, there is variation in the types of connectivity by neighborhood.

San Jose (Figure 4.3) has even greater variation within the city, especially for cell-only subscriptions. These range from 1.7% to a remarkable 40.6% of households across zip codes. Broadband adoption (all types) is lower in some zip codes in San Jose than in the other highly ranked cities. Zip codes there vary from 40.6% to 96.4% with broadband. The lowest end of the broadband subscription continuum in San Jose is more than 20 percentage points below

(a)   (b)

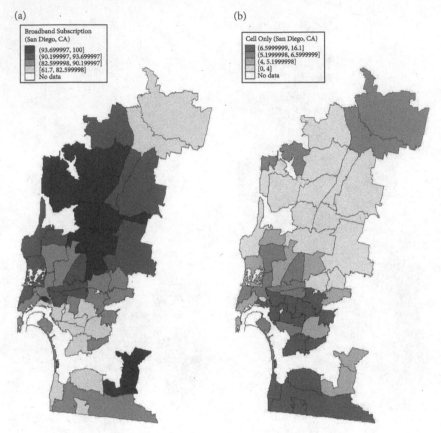

Figure 4.2 San Diego, California (ranked #2 for fixed broadband subscriptions).

the least connected zip codes in the other two top-ranked cities. The same pattern of spatial inequality in types of internet access is evident.

Comparing zip codes in the three lowest-ranked cities also demonstrates a deeper level of disadvantage than citywide figures suggest. This is clearest in Memphis, which ranks #48 for wireline broadband adoption (Figure 4.4). In Memphis, 56.7% of households have fixed broadband subscriptions and 71.8% have some type of broadband, with substantial inequality across neighborhoods. Broadband subscriptions by zip code vary from a low of only 26.4% to a high of 88.9% of households. Cell-only internet use is 15.1% citywide but ranges from 4.76% to a high of 22.2% of households across zip codes. There are neighborhoods where both cell-only internet use and broadband are relatively high on the east

(a)

(b)

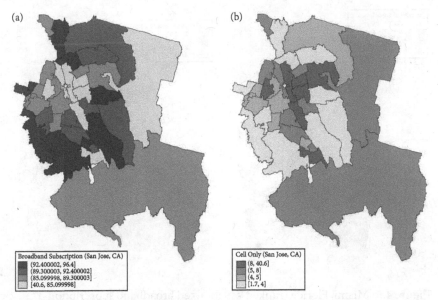

Broadband Subscription (San Jose, CA)
- (92.400002, 96.4]
- (89.300003, 92.400002]
- (85.099998, 89.300003]
- [40.6, 85.099998]

Cell Only (San Jose, CA)
- (8, 40.6]
- (5, 8]
- (4, 5]
- [1.7, 4]

Figure 4.3 San Jose, California (ranked #3 for fixed broadband subscriptions).

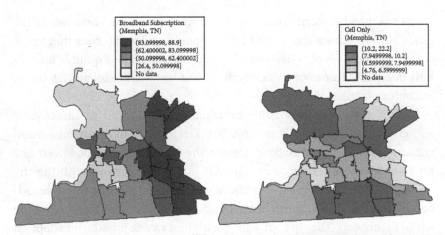

Broadband Subscription
(Memphis, TN)
- (83.099998, 88.9]
- (62.400002, 83.099998]
- (50.099998, 62.400002]
- [26.4, 50.099998]
- No data

Cell Only
(Memphis, TN)
- (10.2, 22.2]
- (7.9499998, 10.2]
- (6.5999999, 7.9499998]
- [4.76, 6.5999999]
- No data

Figure 4.4 Memphis, Tennessee (ranked #48 for fixed broadband subscriptions).

side of the city, as well as a portion in the west where both are in the lowest quintile.

Miami ranks #49 for fixed broadband and #50 for broadband of any type (Figure 4.5). A little over half of Miami residents (54.1%) have fixed broadband. Zip codes in the city vary from a low of 40.8% to 94.2% of households

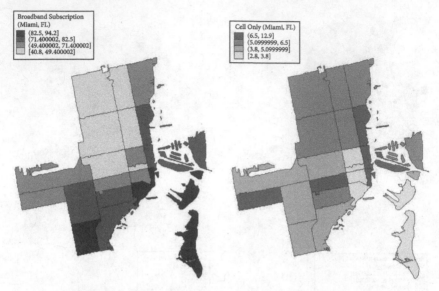

Figure 4.5 Miami, Florida (ranked #49 for fixed broadband subscriptions).

with broadband subscriptions. The lowest rates are on the north side of the city. Cell-only subscriptions are 11.8% citywide, and by zip code they range from 2.8% to 12.9% of households. Zip codes in the highest quintile for cell-only internet use are not necessarily the lowest for fixed broadband adoption, in contrast to other cities.

Detroit ranks #50 in wireline broadband (48.3%) and #1 in cell-only subscriptions (19.2%) (Figure 4.6). But variation in rates for broadband subscriptions is less dramatic in Detroit than in Memphis. The lowest rate for broadband subscriptions by zip code is 46.4% compared with the city average of 67.5% (all types). Zip codes in the lowest quintile for all broadband tend to be in the north central and near west side of the city (on the left in Figure 4.6). The zip code with the highest rate of broadband adoption is at 86.7%. Though cell-only internet subscriptions are particularly high in Detroit, at 19.2%, some areas in the center of the city are in the lowest or second lowest quintile for cell-only internet use as well as for broadband of all types. These zip codes have low connectivity of any kind and demonstrate the deep-rooted disadvantage in some central neighborhoods. There is some mirroring of broadband and cell-only internet use toward the northeastern and western edge of the city.

(a)

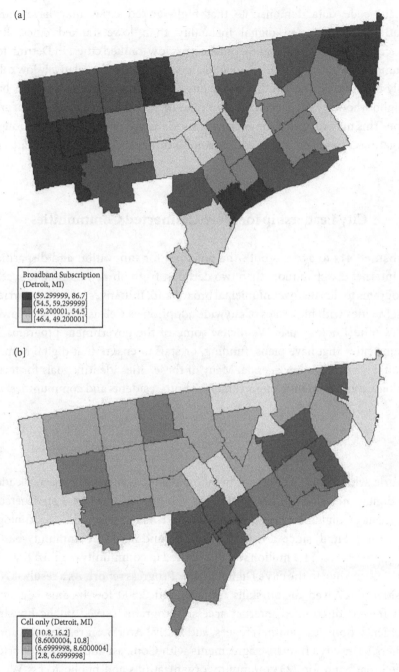

Broadband Subscription
(Detroit, MI)
- ■ (59.299999, 86.7]
- ▨ (54.5, 59.299999]
- ▨ (49.200001, 54.5]
- ▢ [46.4, 49.200001]

(b)

Cell only (Detroit, MI)
- ■ (10.8, 16.2]
- ▨ (8.6000004, 10.8]
- ▨ (6.6999998, 8.6000004]
- ▢ [2.8, 6.6999998]

Figure 4.6 Detroit, Michigan (ranked #50 for fixed broadband subscriptions).

Zip code data demonstrate that high-ranked cities may have more spatially concentrated digital inequality than lower-ranked cities. But concentrations of disadvantage differ across low-ranked cities. In Detroit, for example, low rates of fixed broadband are accompanied by relatively low cell-only use in large areas of the central city and near southwest. Memphis has neighborhoods with particularly low rates of adoption for broadband of any type. This may require different strategies for addressing inequality, through targeting certain communities in Memphis or broader citywide programs in Detroit.

City Leadership for More Connected Communities

Urban efforts to address both the potential for innovation and disparities in internet use span more than two decades, from library and public access programs to debates over municipal broadband. Initiatives have been carried out in cities with high rates of citywide adoption as well as places with lower rates of technology use. We review some of the government programs in major cities that have plans, funding, or staff to ensure that digital human capital is on the policy agenda. Many of these cities identify goals for technology use as economic opportunity for both residents and communities.

Seattle

Seattle, which ranks #1 in home broadband adoption, has long been a leader in digital inclusion efforts. Seattle was the first city to hire a staff person to manage digital inclusion initiatives and has sponsored a Technology Matching Fund since 1998 to provide funding for community-based programs.[6] Over $1.4 million was distributed to community organizations in 2018, according to the city's Digital Equity Progress report. As a result, 4,700 residents received digital skills training and 2,500 low-income residents got free or discounted internet access.[7] Programs served public housing residents, homeless youth, refugees, and Native American residents, among others.[8] The city's franchise agreements with Comcast and Wave supported free broadband for 223 community organizations and public access Wi-Fi in 70 locations. The city created a Digital Equity Plan in 2015, and city policies have been informed by five citywide surveys since 2000. In a forward to Seattle's 2018 Technology Access and Adoption Study, Mayor Jennifer Durkin declared that "Being the city that invents the future means leaving nobody behind."

Austin

Austin's city council approved its Digital Inclusion Strategic Plan in 2014.[9,10] An online dashboard reports progress on specific action steps subsequently identified in a business plan created in 2016.[11] The plans are implemented through partnerships; the Digital Empowerment Community of Austin (DECA) is a network of nonprofits and schools that collaborates with the city[12] and the nonprofit Austin Free-Net manages the city's public access labs and delivers digital literacy training.[13] The city has an agreement with Google Fiber to provide free gigabit broadband at city hall, the central library, and up to 100 sites associated with public or nonprofit organizations.[14] Austin also funds the Grant for Technology Opportunities Program. In 2018, the initiative provided $200,000 through nine grants supporting community organizations.[15]

Portland

Portland's Office for Community Technology coordinates the city's participation in digital inclusion initiatives and has served in this role since 2010. It staffs the Digital Inclusion Network, a cross-sectoral group of stakeholders that formed in 2014 and includes the city of Portland, the Multnomah County Public Library, nonprofits, businesses, higher education, and media organizations.[16] The city council adopted the Digital Equity Action Plan in 2016,[17] and in 2018 the Digital Inclusion Summit on Economic Opportunity attracted representatives from 69 organizations to promote "building a digitally connected, prosperous community."[18] Portland's Office for Community Technology participates in the Smart Cities Steering Committee to "ensure digital inclusion and equity values are central in the development of Portland's Smart Cities vision, policies and projects."[19]

Boston

According to Boston's website for Broadband and Digital Equity, the city wants "residents to get the skills they need to find success in the 21st century. This effort will help us make sure Boston becomes a more fair and innovative City."[20] In 2015, Boston was the second US city to hire a full-time staff person to promote digital inclusion (Siefer 2016), and Boston established the Digital Equity Fund in 2017. The city provides Wicked Free Wi-Fi in libraries and 20 other sites, along with programs in libraries and community centers. It supports most of the $1 million annual budget for Tech Goes Home (Johnston 2016), which has trained 30,000 people at over 300 school and community sites and has distributed 20,000 computers. The Tech Goes Home program in the schools provides classes for parents or caregivers as well as students (Schartman-Cycyk et al. 2019).

Louisville

According to the Louisville Metro Plan, the southwest and west sides are "fiber deserts" with low rates of broadband adoption and high unemployment. The plan outlines goals for extending public access and broadband infrastructure, providing training, and partnering with nonprofits to provide affordable hardware and software. The expected arrival of Google Fiber led to planning for a gigabit experience center in the Russell neighborhood on the west end, to transform the community into a smart neighborhood. More generally, Louisville's plan stresses economic opportunity and human capital; Metro's digital equity goals are jobs, education, and compassion (defined as inclusive access).

Philadelphia

Philadelphia has the highest poverty rate among the 10 largest cities (Li and Sussman 2018) and has a long history of technology efforts, including attempts to create a municipal broadband network that faced challenges from incumbent internet providers and the state legislature (Abraham 2015). Today the city manages a network of community-based organizations to deliver public access and training through the KEYSPOT program (Abraham

2015; Li and Sussman 2018). Philadelphia sponsors a Digital Literacy Alliance that includes nonprofits, universities, and internet service providers. Franchise fees from Comcast and Verizon support grants to community organizations, and the fund was approximately $700,000 in 2019.[21] Through a grant from the Knight Foundation, the city has created a SmartCityPHL Roadmap, with a plan for 100 digital kiosks that will offer free public Wi-Fi. The 55-inch screens will display information on public services, local events, and emergency messages. A tablet built into the kiosks will offer information on city services, maps, and directions (Descant 2019).

Detroit

In April 2019, the city of Detroit hired its first director of digital inclusion, to tackle the low rates of broadband adoption in the city. The position is under the director of the Department of Innovation and Technology (Government Technology 2019). The city now has hotspot lending at libraries, a Wi-Fi asset map, low-cost broadband options, computer labs at churches, storefronts for low-cost IT, and classes run by partner organizations.[22] Rocket Mortgage is collaborating with the city and United Way to provide a 313 Connect Fund to improve internet access (Balboa 2020).[23]

Cities have created partnerships with nonprofit organizations, funded neighborhood efforts, and hired dedicated staff to institutionalize commitments to increasing the community's digital human capital. They have formulated digital inclusion or digital equity plans in Austin (2014), Seattle (2015), Portland (2016), and Louisville (2017). Boston hired a staff person for digital inclusion in 2015, and Detroit followed suit in 2019. Smart city investments are one factor that has increased the need for digital inclusion efforts (Horrigan 2019).

These are but a few examples where city governments have exercised leadership for digital inclusion, through partnerships with community-based organizations, foundations, and the private sector. Funding for local actions may be derived from cable franchise fees or charges for use of public right-of-way, as well as local budgets. In the examples cited in the case studies, Google Fiber, Microsoft, Comcast, and other corporations have contributed funds or partnered with cities or community organizations, though efforts for municipal broadband have involved conflict with providers in the past. Foundations have also supported city and nonprofit programs. New

Orleans held a Digital Equity Challenge with support from the Rockefeller Foundation (LaRose 2018),[24] and foundations have played an important role in Philadelphia, Detroit, and the program we discuss next.

Chicago's Smart Communities and Neighborhood-Level Change

Although residence in poor neighborhoods magnifies barriers to technology use (Mossberger, Tolbert, Bowen, and Jimenez 2012), can neighborhoods instead be a positive force for change? Early studies demonstrated that computer purchases are more likely in geographic areas where a high proportion of households already own computers (Goolsbee and Klenow 2002). Local spillovers and learning from others might be expected for broadband adoption too. This may be especially true in low-income communities, where there is high internet use outside the home, including at the homes of friends and relatives. This outside use may encourage learning, and the example of friends and family may influence home adoption (Mossberger, Kaplan, and Gilbert 2008; Mossberger et al. 2003).

The Smart Communities program, implemented between 2010 and 2013, aimed to create a culture of digital excellence, or IT use, in 9 of Chicago's 77 community areas, or neighborhoods (LISC Chicago 2009, 3; see also Mayor's Advisory Council on Closing the Digital Divide 2007). The city of Chicago received a $7 million federal stimulus grant in 2009 for the Smart Communities program. It was implemented by five community-based lead agencies and the Chicago Local Initiatives Support Corporation (LISC). An evaluation of the Chicago Smart Communities program was supported by the John D. and Catherine T. MacArthur Foundation and the Partnership for a Connected Illinois, making possible the collection of unique neighborhood-level data on broadband subscriptions, barriers to use, and activities online (see Mossberger, Tolbert, and Anderson 2014).

The evaluation, discussed in this final section, provides rare insight into the effectiveness of digital inclusion programs for promoting change on a neighborhood scale. The treated neighborhoods were low and moderate income and at least 75% African American and Latino. Three thousand residents participated in training along with 500 small businesses. The program brought together multiple training and outreach efforts delivered through existing community organizations: technology skills training in English and Spanish at

FamilyNet Centers, digital summer jobs, training and technical assistance for internet use by small businesses, Civic 2.0 classes for neighborhood groups, and digital media programs for youth. The Smart Communities, however, did not provide internet access. Outreach encouraging purchase of a broadband subscription was conducted through tech organizers, neighborhood portals, and advertising on buses and transit shelters. In fall 2011, midway through the program, Comcast began to offer discounted broadband to households in Chicago with children who were receiving free or discounted school lunch. Here we discuss the main findings of the evaluation.

Measuring Broadband Subscriptions for Chicago Neighborhoods over Time

The program evaluation was based on unique neighborhood-level data for Chicago's 77 official community areas measuring broadband subscriptions over time, as well as online activities (Mossberger, Tolbert, and Anderson 2014). It used a pre- and postdesign to measure whether the Smart Community neighborhoods experienced different rates of change in technology access and use compared to other Chicago community areas. Estimates for broadband subscriptions, including internet use anywhere, home adoption, and activities online, were based on citywide random digit-dialed telephone surveys conducted in 2008 (3,500 respondents), 2011 (2,500 respondents), and 2013 (2,400 respondents) by Rutgers's Eagleton Institute using landline and mobile.[25] The surveys were conducted in English and Spanish and included respondents from all 77 community areas. Similar questions on internet access and activities online were asked in all three years, allowing for a comparison over time.

To generalize from a small sample in each community area to an entire neighborhood can be problematic and lead to bias. To overcome this problem, hierarchical linear modeling was used to estimate internet adoption and use for Chicago's 77 neighborhoods at three points in time (2008, 2011, and 2013).[26] We use random intercept multilevel statistical modeling with poststratification weights (a form of statistical simulation) to generate geographic estimates of broadband access and online activities for neighborhoods in Chicago. This is the same methodology used to create the time series estimates of broadband subscriptions reported in other chapters of this book.

Respondents in the three surveys were asked to identify their cross-streets. This information was used to geocode each participant in a census tract. The survey data was merged with aggregate-level census tract information from the ACS for the appropriate citywide survey (2008, 2011, or 2013). The statistical models are based on data that combines individual and aggregate variables. The results are simulated predictions of broadband subscriptions and online activities for each of Chicago's 77 community areas over time. Leveraging the neighborhood-level data provides more accurate and representative estimates than could be obtained from the individual-level data alone.

Comparing Rates of Change from 2008 to 2013 for Treated and Nontreated Neighborhoods

With the data measuring broadband subscriptions and use for Chicago neighborhoods, multivariate regression analysis was used to compare change in the Smart Communities compared to nontreated neighborhoods over the five-year period (see Mossberger, Tolbert, and Anderson 2014). The primary predictor of interest was a binary variable that measured whether the neighborhood had Smart Communities programs or not. The outcome or dependent variable was created by taking the difference in our estimates from 2013 minus 2008 (Mossberger and Tolbert 2009); it measures change in broadband subscriptions or online activities by neighborhood over the five-year period. To account for other factors that might lead to changes in technology use for neighborhoods, the models controlled for changes in population, percentage in poverty, high school graduates, and the race, ethnicity, and age of the population (2008 and 2012 ACS).[27] The models control for change in the neighborhoods in terms of socioeconomic status, racial and ethnic diversity, and age—possible explanations for changes in neighborhood broadband use other than the Smart Communities program.

Results: Higher Rates of Broadband Subscriptions from 2008 to 2013

Controlling for changing demographic and economic conditions in the neighborhoods, residents of Smart Communities neighborhoods had a higher rate of increase in internet use in any location than other areas of the city.[28] Figure 4.7 (left) graphs the predicted rate of change in internet use

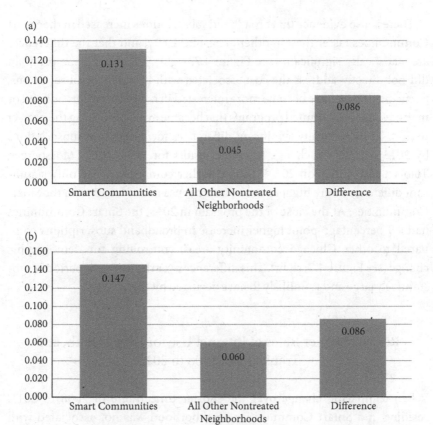

Figure 4.7 Predicted change in internet use (a) and change in broadband subscriptions (b) from 2008 to 2013.

Note: Estimates based on multivariate statistical models with other neighborhood-level factors (change in poverty rates, demographic factors, etc.), held constant at mean values. See Mossberger, Tolbert, and Anderson (2014).

at any location in the Smart Communities compared to other community areas from 2008 to 2013. Holding change in all other demographic and economic factors constant at mean values, internet use in any location increased by 4.5% in nontreated communities and 13% in the Smart Communities. This means there was a 9-percentage-point higher increase in internet use in the Smart Communities compared to other Chicago neighborhoods. This change includes residents who do not have a broadband subscription but who use the internet on smartphones or who use public access sites like libraries, the homes of friends and relatives, or other internet connections at places outside the home.

There is also evidence that broadband subscriptions increased in the Smart Communities faster than in other areas of the city, and that the differences are statistically significant (see Figure 4.7, right).[29] Although the program did not directly address the cost associated with broadband subscriptions, participants may have become more interested in having home broadband or may have used discounted programs that became available during the project period. The statistically significant differences for broadband subscriptions by 2013 represent a change from the results for 2008–2011 (Mossberger, Tolbert, and Anderson 2012). In that earlier comparison, the only significant differences were higher rates of internet use in any location in the Smart Communities. At the close of the program in 2013, the Smart Communities had a 9-percentage-point higher increase in broadband subscriptions compared to other Chicago community areas, controlling for demographic change (see Figure 4.7, right). This difference over the five-year period is substantively large and is unlikely to have occurred by chance.

Results: Higher Rates of Internet Use for Jobs, Health, and Transit from 2008 to 2013

When comparing change over the shorter time frame of 2008 to 2011, residing in a Smart Communities neighborhood was not associated with a statistically higher rate of change in any online activities. But when comparing change over the five-year period (2008 to 2013), the Smart Communities neighborhoods had significantly higher rates of change for using digital information on jobs, health, and mass transit. By 2013, the rate of increase for online job search was 11 percentage points higher in the Smart Communities than other Chicago neighborhoods (see Figure 4.8, left).[30] Change in use of the internet for health information was 17 percentage points in the Smart Communities, whereas the increase was only 6.6 percentage points in other neighborhoods, a 10-percentage-point difference (see Figure 4.8, right). Similarly, the increase in use of digital information for mass transit was 17 percentage points in the Smart Communities and 6 percentage points in other Chicago neighborhoods, an 11-percentage-point difference comparing the treated and nontreated neighborhoods (Figure 4.9). This suggests a measurable effect not only in increasing internet access and broadband use but also in a range of important economic and social activities online. The Smart Communities experienced a 10%

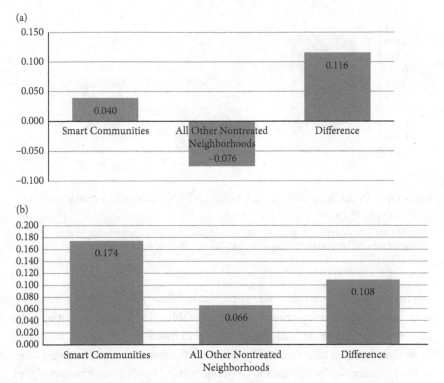

Figure 4.8 Predicted change in online search for a job (a) and health information from 2008 to 2013 (b).

Note: Estimates based on multivariate statistical models with other neighborhood-level factors (change in poverty rates, demographic factors, etc.), held constant at mean values.

to 11% higher increase in online search activities for health information and transportation over five years, compared to similarly situated Chicago neighborhoods.

Other online activities included in the surveys did not demonstrate statistically significant differences from other Chicago neighborhoods in rates of change over the five-year period: use of the internet for political information, for an online class or training, for government information (any government), or the city of Chicago website.

Summing Up and Considering Other Evidence

Analysis of the Chicago data indicates that the Smart Communities experienced statistically significant, higher rates of increase in internet use

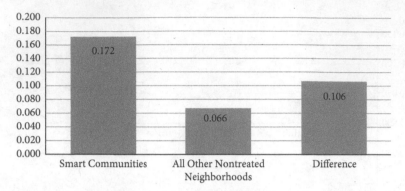

Figure 4.9 Predicted change in online search for transportation information from 2008 to 2013.

Note: Estimates based on multivariate statistical models with other neighborhood-level factors (change in poverty rates, demographic factors, etc.), held constant at mean values.

anywhere, fixed broadband adoption, and online job, health information, and mass transit information search. Notably, the online activities with higher rates of increase, including health and mobility, contribute to human capital and economic opportunity. These differences are substantively large as well as statistically significant, between 9 and 11 percentage points higher than in nontreated neighborhoods. Designation as a Smart Community was not randomly assigned, so the evaluation was quasi-experimental.

While it is impossible to rule out all explanations for the change in internet use other than the Smart Communities efforts, the statistical controls used here are useful for eliminating known challenges such as demographic change. Additionally, estimating access at three points in time provided insights on the process of change. The change observed in the 2013 data compared to 2011 fits with Horrigan's (2015) evaluation of Internet Essentials participants, who deepened their engagement with the internet over time.

The community-level comparisons discussed here can be triangulated with other evidence, including surveys of FamilyNet and Civic 2.0 program participants in the Smart Communities evaluation (Mossberger, Feeney, and Li 2014; Mossberger, Benoit Bryan, and Brown 2014). Results from the FamilyNet surveys suggest that informal learning and resource sharing may have been one factor helping to spread the benefits of classes beyond trainees. One-third of the FamilyNet respondents reported helping others to use the internet, and half of those they helped lived in the neighborhood (Mossberger, Feeney, and Li 2014). Along with outreach by tech organizers and the advertising campaign, resource sharing may have helped to create some spillover effects within these communities. The Smart Communities offered classes for community organizations on how to use the internet to research issues online and to access government services. Survey results from these participants and interviews with community organizations indicate that neighborhood groups may have participated in creating a spillover effect as well (Mossberger, Benoit Bryan, and Brown 2014).

With the ACS data now available by census tract for 2017, it is possible to aggregate the five-year estimates by Chicago community area and to see how the Smart Communities neighborhoods have fared four years later (Figure 4.10). There is variation across these neighborhoods, but seven of the nine are in the top two categories for broadband subscriptions (shaded darker) when cell-only internet use is taken into account, shown in Figure 4.10 (right). Like many low-income communities of color, they are less likely to have high rates for fixed broadband subscriptions, shown in Figure 4.10 (left). Gains made by 2013 were not equally maintained and built upon afterward. Analysis of the 2017 ACS data shows that the Smart Communities neighborhoods that continued to have higher increases in broadband subscriptions by 2017 also had more residents with at least a two-year college degree (Tolbert et al. forthcoming). The Smart Communities intervention brought about neighborhood-level change in digital human capital during the two-year program, but this was most likely to be sustained in the long run when other forms of human capital were present in communities as well. This was a short-term program that ended in 2013 with the lapse in federal funding, but it is possible that renewed efforts, such as Cook County's new Council on Digital Equity (CODE), may further improve digital outcomes in the city with more long-term attention to affordability and skills (Hinton 2020).

The Smart Communities demonstrated that change is possible, and there has been a dearth of evidence on whether programs encouraging broadband adoption and use are effective in promoting community-level change. Most evaluations of broadband outreach and training programs examine implementation rather than community-level outcomes (Hauge and Prieger 2010, 2015). The evidence that exists in prior studies shows mixed results. ConnectKentucky's education and training program had incremental effects on broadband adoption but did not increase economic activities using the internet (LaRose et al. 2011). A later examination of ConnectedNation's outreach and training in five states concluded that there was no significant effect (Manlove and Whitacre 2019). Nationally, spending on federal stimulus

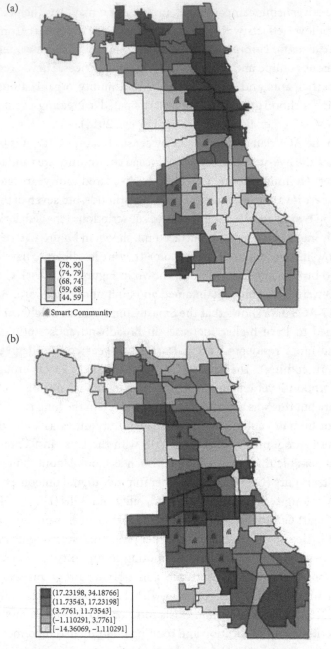

(a)

(78, 90]
(74, 79]
(68, 74]
(59, 68]
[44, 59]

Smart Community

(b)

(17.23198, 34.18766]
(11.73543, 17.23198]
(3.7761, 11.73543]
(−1.110291, 3.7761]
[−14.36069, −1.110291]

Figure 4.10 Broadband subscriptions in 2017 by Chicago community area, including cell phone only (a) and omitting cell phone only (b).

*Smart Communities marked by icon.

Source: Tolbert et al. (forthcoming).

programs did not necessarily lead to better outcomes at the county level, as measured by the FCC's categorical adoption data. Clearly some places did experience improvements, however, and researchers suggested that more evaluation of individual programs was needed (Hauge and Prieger 2015). The Chicago Smart Communities program is an example of these stimulus efforts, and one in which we are able to use more precise data than in the Hauge and Prieger study, as well as comparisons over time with nontreated Chicago neighborhoods.

Policy interventions are often place based, and geographic measures of change offer community residents and policymakers useful data for understanding whether programs are effective and how to target and invest scarce resources going forward. The role of place is also important for research and policy addressing digital human capital and broadband use, including tracking longer-term outcomes for community development and resident opportunity.

Conclusion: Smart Cities and Equity— A Vision for the Future?

Cities are at the heart of innovation globally. The smart cities movement, to echo the quote at the beginning of the chapter, is focused on creating a vision for the future through broadband applications. For technology to truly benefit the public interest, it must foster widespread digital human capital—for access to information and economic opportunity, and for participation in society online.

Yet there are wide differences in broadband adoption across cities. Home broadband subscriptions range from 85% of households in Seattle to only 48% in Detroit. Even if smartphones are included, one-third of households in Detroit and Miami have no personal broadband access at all. This contrasts with the 10 highest-ranked cities, where broadband connections that include smartphones approach or exceed 90% of households. Racial and ethnic disparities are a feature of all cities, though the gaps vary. Connectivity, including cell phones, is only 50% to 70% for African American and Latino households in Detroit, Miami, Memphis, Milwaukee, Houston, and Dallas. There are neighborhoods with low rates of adoption even in the highest-ranked cities, and in Memphis there are zip codes where only 26% of households have fixed broadband. Maps of wireline broadband adoption and cell-only internet use nearly mirror each other, demonstrating that cities are indeed developing new fault lines, reinforcing segregation and concentrated poverty. Higher rates of smartphone-only internet use by low-income African Americans and Latinos have not erased technology disparities. Affordable fixed broadband is still needed for full connectivity, for developing skills, for engaging in a wide range of activities online, and for access to opportunity regardless of zip code. Those who are fully connected have both fixed broadband and mobile.

Local governments have been praised in scholarly and policy circles for collaborative, cross-sectoral leadership for creative and pragmatic problem solving (Katz and Nowak 2017; Barber 2013; Goldsmith and Crawford 2014). As demonstrated in the profiles in this chapter, some cities are indeed leading efforts to create more digitally inclusive communities, though equity is not always a priority in smart city planning (Horrigan 2019). Some cities have dedicated staff, passed official inclusion plans, and partnered with companies, nonprofits, and community groups. These local governments have acted as conveners, building multisector coalitions. As the Smart Communities evaluation demonstrated, neighborhood-level change is indeed possible and can lay the foundation for broader community building and the cities of the future.

Appendix Table 4.1 Ranking Cities by Broadband Subscriptions[*]

Cities	Rank Fixed Broadband	%	Rank Fixed + Cell Only	%	Rank Cell Phone Only	%
Seattle, WA	1	85.2	4	91.3	50	6.1
San Diego, CA	2	83.9	2	91.5	46	7.6
San Jose, CA	3	83.9	3	91.4	47	7.5
Raleigh, NC	4	83.7	1	92.4	43	8.7
Virginia Beach, VA	5	83.6	5	90.9	48	7.3
Colorado Springs, CO	6	81.3	6	90.4	38	9.1
San Francisco, CA	7	80.9	11	87.8	49	6.9
Portland, OR	8	80.6	7	89.7	39	9.1
Denver, CO	9	78.9	14	87.4	44	8.5
Austin, TX	10	78.1	9	89.4	27	11.3
Sacramento, CA	11	77.3	8	89.5	23	12.2
Boston, MA	12	76.7	18	85.5	42	8.8
Oakland, CA	13	76.6	17	85.6	40	9
Charlotte, NC	14	76.6	12	87.8	29	11.2

Appendix Table 4.1 *Continued*

Cities	Rank Fixed Broadband	%	Rank Fixed + Cell Only	%	Rank Cell Phone Only	%
Columbus, OH	15	76.5	13	87.5	30	11
Minneapolis, MN	16	75.7	29	84	45	8.3
Long Beach, CA	17	74.8	28	84.1	37	9.3
Omaha, NB	18	74.3	25	84.6	33	10.3
Arlington, TX	19	74.1	19	85.3	28	11.2
Los Angeles, CA	20	74	24	84.6	31	10.6
Washington, DC	21	73.7	32	82.7	41	9
Nashville, TN	22	73.3	16	85.9	21	12.6
Wichita, KS	23	73.3	30	83.5	36	10.2
Albuquerque, NM	24	72.7	27	84.4	25	11.7
Mesa, AZ	25	72.4	10	88.8	4	16.4
Atlanta, GA	26	72.3	21	85	20	12.7
New York, NY	27	72.2	33	82.4	35	10.2
Fort Worth, TX	28	71.3	15	86.4	10	15.1
Louisville, KY	29	71	20	85.2	14	14.2
Phoenix, AZ	30	71	26	84.6	17	13.6
Tucson, AZ	31	70.1	22	84.8	11	14.7
Kansas City, MO	32	69.7	34	82.3	22	12.6
Las Vegas, NV	33	69	43	79.3	32	10.3
Jacksonville, FL	34	68.9	35	82.2	18	13.3
Oklahoma City, OK	35	68.6	23	84.7	5	16.1
Chicago, IL	36	68.5	41	79.9	26	11.4
Fresno, CA	37	68.4	36	82	16	13.6
Indianapolis, IN	38	67.1	38	80.8	15	13.7
San Antonio, TX	39	66.8	42	79.7	19	12.9
Tulsa, OK	40	66.7	37	81.1	13	14.4
Houston, TX	41	66.3	31	83.4	3	17.1
Dallas, TX	42	65	40	80.3	7	15.3
El Paso, TX	43	63.3	39	80.8	2	17.5
Baltimore, MD	44	61.6	44	76.2	12	14.6
Philadelphia, PA	45	61.4	48	71.6	34	10.2
New Orleans, LA	46	60	45	75.1	9	15.1
Milwaukee, WI	47	57.7	46	73.1	6	15.4
Memphis, TN	48	56.7	47	71.8	8	15.1
Miami, FL	49	54.1	50	65.9	24	11.8
Detroit, MI	50	48.3	49	67.5	1	19.2

*Shaded cells indicate top and bottom 10 cities by broadband subscription rates.

Appendix Table 4.2 Cities by Broadband Subscriptions by Racial and Ethnic Groups*

Cities	Rank Broadband	% Broadband Black	% Broadband Hispanic	% Broadband Asian	% Broadband White	% Broadband
Seattle, WA	1	75.7	85.1	87.9	92.9	91.3
San Diego, CA	2	83.3	83	93.6	93.2	91.5
San Jose, CA	3	87.4	85.7	94.6	92.6	91.4
Raleigh, NC	4	80.5	78.1	92.4	93	92.4
Virginia Beach, VA	5	82	90.2	92.3	91.3	90.9
Colorado Springs, CO	6	86.7	86.5	92.5	91.9	90.4
San Francisco, CA	7	73.4	85.4	87.9	92.4	87.8
Portland, OR	8	78.7	84.8	88	90.3	89.7
Denver, CO	9	76.1	73.9	89.5	89.4	87.4
Austin, TX	10	75	72	94.1	92.1	89.4
Sacramento, CA	11	76	78.9	88.1	86.7	89.5
Boston, MA	12	78.2	80.8	86.6	90.1	85.5
Oakland, CA	13	74.2	74.4	83.4	92.6	85.6
Charlotte, NC	14	78.6	74.5	92.8	92.4	87.8
Columbus, OH	15	75.4	75.4	90.4	88.4	87.5
Minneapolis, MN	16	65.6	74.7	85.6	89	84
Long Beach, CA	17	75.1	76.6	85.9	89.7	84.1
Omaha, NB	18	66.3	67.1	86.3	87.9	84.6
Arlington, TX	19	76.8	73.3	86.1	88.7	85.3
Los Angeles, CA	20	71.6	73.9	89	89.8	84.6
Washington, DC	21	65.7	77	90.5	95.6	82.7

City	Rank					
Wichita, KS	23	61.7	68.2	91.9	84.1	83.5
Albuquerque, NM	24	74.2	75.2	90.2	86.3	84.4
Mesa, AZ	25	82	75.6	92.9	88.4	88.8
Atlanta, GA	26	66	73.4	92.4	92	85
New York, NY	27	77.5	76.8	87.7	83.8	82.4
Fort Worth, TX	28	68.9	70.6	90	90.7	86.4
Louisville, KY	29	69.3	77.1	86.9	84.9	85.2
Phoenix, AZ	30	72.2	68.1	90.8	87.5	84.6
Tucson, AZ	31	76.8	81.1	83.5	85	84.8
Kansas City, MO	32	67	72.3	82.8	87.2	82.3
Las Vegas, NV	33	67.9	73.1	88.7	86.4	79.3
Jacksonville, FL	34	71.6	80	89.9	86	82.2
Oklahoma City, OK	35	68.2	74.9	89	84.9	84.7
Chicago, IL	36	64.7	73.7	87.5	87	79.9
Fresno, CA	37	68.4	74.7	83.2	85.9	82
Indianapolis, IN	38	68.4	72.2	87.2	83.6	80.8
San Antonio, TX	39	73.2	72.6	91.3	87.1	79.7
Tulsa, OK	40	65.7	71.1	89.1	83.7	81.1
Houston, TX	41	69.1	68.5	90	89.6	83.4
Dallas, TX	42	59.3	64.3	87.9	87.7	80.3
El Paso, TX	43	83	74.9	88.6	87.7	80.8
Baltimore, MD	44	68.3	72.3	89.9	85.1	76.2
Philadelphia, PA	45	68.8	72.4	86.3	82.4	71.6
New Orleans, LA	46	59.7	73.4	81.4	88	75.1
Milwaukee, WI	47	64.7	68	84	81.3	73.1
Memphis, TN	48	59.5	55	84.7	83.5	71.8
Miami, FL	49	51.9	68.1	78	86.1	65.9
Detroit, MI	50	57.2	57.7	72.2	71.5	67.5

*Shaded cells indicate top and bottom 10 cities by broadband subscription rates.

5

States

The Innovation Environment

All 50 states have some type of broadband policy, but programs vary widely in their goals, funding, and implementation. States regulate who can provide broadband, with 22 either prohibiting or imposing restrictions on local government participation in municipal networks (Broadband Now 2020), and others that enable public provision through entities such as utility districts (Maine) and electric cooperatives (Missouri, West Virginia). To encourage investment by businesses, some states (Wisconsin, Georgia, Indiana, and Tennessee) certify local communities as broadband ready (Pew Charitable Trusts 2019). Three-quarters of the states have a dedicated broadband office, and 27 states have a designated fund for grants, although not all of them are funded (Stauffer et al. 2020). Most states focus on deployment, especially in rural areas, though some, like California, address broadband use in urban and rural communities. As part of a settlement allowing mergers for major telecommunications firms, the state established a $60 million quasi-governmental California Emerging Technology Fund (CETF n.d.). The organization makes grants to local governments and nonprofit organizations to advance broadband awareness, adoption, and skills (Stauffer et al. 2020, 11–12).

Broadband programs are just one example of how policies vary across the states. As with many other policies in the federal system, states have substantial autonomy to create their own programs and to determine the rules of the game for their counties and cities. States set the context for innovation and prosperity at the local level, not only through broadband programs but also across many policies affecting communities and economic development. With the ability to formulate different approaches, the 50 states have long been viewed as "laboratories of democracy"[1] that promote experimentation and learning. At the core of this process is information—how policy ideas, research, and experience circulate and get used by state decision makers (Mossberger 2000, 6). This information has long spread across

Choosing the Future. Karen Mossberger, Caroline J. Tolbert, and Scott J. LaCombe, Oxford University Press. © Oxford University Press 2021. DOI: 10.1093/oso/9780197585757.003.0005

state boundaries through federal agencies, professional associations, media, think tanks, university researchers, political parties, advocacy groups, and informal contacts in other states (Mossberger 2000, 100–101; Mintrom 2000; Balla 2001). With the advent of the internet, it is immeasurably easier and quicker to access information from more sources, near and far, expanding and accelerating these networks of information and innovation.

This chapter explores the effects of digital human capital on state innovation, asking whether states with higher levels of broadband subscriptions in the population are more likely to be at the leading edge of the spread of new policy ideas. Are states with more widespread broadband use more likely to be earlier adopters of new policies?[2] Measuring the economic outcomes of digital human capital is more meaningful in counties and metropolitan areas, as larger and more diverse states include multiple regional economies, urban and rural. But states are ideal sites for examining the impact of inclusive broadband as an infrastructure for information and policy innovation.

As we argued in Chapter 1, digital human capital should encourage the development of information networks within and across geographies. This should promote the flow of policy ideas, evidence, and opinion within states. Through digital technologies, members of the public have easier access to information and opportunities to communicate their preferences to legislators. Interest groups have more potential sources of information and instant connections with others around the nation. Local governments can more easily lobby for changes at the state level. Contacts with other states, professional associations, and federal agencies are enhanced with online communications in both state administrations and legislatures. Digital media and communications (news, email and texting, podcasts, listservs, social media, streaming video, and more) have fundamentally altered how information is disseminated in recent decades—for both policymakers and the public.

In keeping with prior research in the field, we define policy innovation as early adoption of a policy idea even if it originated elsewhere (Eyestone 1977). Our analysis uses the State Policy Innovation and Diffusion Database (Boehmke et al. 2019), a comprehensive dataset on state policy adoptions, to evaluate how the growth in broadband subscriptions over almost two decades has affected the propensity for states to embrace new policies. The database contains hundreds of policies accumulated over the past century, in areas such as education, transportation, workforce development, taxes, and more.

In the next section of this chapter we review the literature on policy innovation and diffusion in the states. Information has played an important part in theories of policy innovation, and broadband data allow us to measure information networks more directly than in the past. We find that broadband growth over time does indeed predict more policy innovation. Prior research has shown that policies often spread from neighboring states, but by introducing the growth of broadband over time and changes across hundreds of policies, we see a decline in these regional influences, pointing to more national networks of information and innovation. We consider the implications of these findings for broader questions about digital human capital and the transformative role of broadband across policy areas.

Patterns of Diffusion and Innovation

Early studies of policy innovation examined patterns of state diffusion and asked whether certain states were more innovative and whether patterns of emulation and information were regional or national (Walker 1969; Gray 1973; Savage 1985). Berry and Berry's (1990) introduction of event history analysis opened a new wave of interest in these topics as researchers could use internal characteristics of states to model innovation and external determinants to identify influences in diffusion networks. Recently a number of studies have taken a large-N approach to measuring policy diffusion, often using dozens or hundreds of policies to systematically identify trends and patterns (Boushey 2010; Boehmke and Skinner 2012; Kreitzer 2015; Karch et al. 2016; Boehmke et al. 2019), and we use this strategy as well. This has enabled researchers to identify consistent patterns across policies and study more systematically what drives policy diffusion and innovation as a general phenomenon.

Despite the variety in methodological and theoretical approaches to understanding how policies spread, information exchanges play a central role in policy diffusion (Boushey 2010; Nicholson-Crotty and Carley 2018). This is typically not measured directly but rather through the ways that policies spread, across neighboring states in a region or through other types of networks. Whether competing, learning, or imitating (Shipan and Volden 2008), states use information about what other states are doing when they adopt new policies.

We focus on two recent dynamics that have been observed by those who study state policy diffusion. First, states have become more innovative over time (Boehmke and Skinner 2012; Boehmke et al. 2020). Whether using Walker's (1969) original innovativeness score or a more recent "rate score" (Boehmke and Skinner 2012), states have experienced a surge in the number of new policies being adopted beginning in the 1990s. This growth in policy innovation mirrors that of the Great Depression, when states quickly adopted a host of alternatives to address the severe economic disruptions and challenges of their time. However, the cause of recent spikes in the number of policy adoptions is less clear. Boushey (2010, 173) suggests that "mass media and communications technology has acted as an important trigger" for policy diffusion, and with new data on broadband subscriptions over nearly two decades, we are able to test for the first time whether digital technology has in fact contributed to this dynamic, increasing the availability of information.

The second aspect of diffusion that we examine is the role of contiguous states in the spread of policies. For decades, scholars have found that states are more likely to adopt policies when neighboring states have already done so (Walker 1969; Gray 1973; Berry and Baybeck 2005). Walker (1969) argued that contiguity plays an important role in diffusion because it facilitates information flows for policymakers to learn about new ideas and receive other relevant cues about whether they should adopt a policy. Proximity means states can easily observe what policies neighboring states are adopting, and citizens near state borders are likely exposed to neighboring state policy changes (Pacheco 2012). Geographic proximity suggests that the cost of getting information about neighboring state policies is lower than learning about state policies hundreds or thousands of miles away; neighboring policies may also provide effective solutions if problems are similar within the region.

Yet some recent studies have found that geographic influence has waned in the last few decades (Mallinson 2021). Instead, Desmarais et al. (2015) leveraged policy adoptions across hundreds of policy areas over time to identify "latent diffusion ties," or pathways new policies persistently take; in so doing the research identifies a state policy network with certain states as likely sources for future policy adoptions. The internet may not only increase the speed at which ideas circulate but also facilitate a greater diversity of information sources, for both citizens and policymakers.

Alternatives, Agendas, and the Spread of Information

The literature on policy diffusion has generally asked where legislators or administrators turn to find policy prescriptions for emerging problems. Ideas considered by policymakers may come from a variety of sources—from interest groups, media, think tanks, professional networks, and informal contacts with other states (Mintrom and Vergari 1998; Mossberger 2000; Balla 2001; Boushey 2010; Garret and Jansa 2015). Information is not just constrained to learning about policies with successful outcomes such as higher economic growth or increased approval from voters. State legislatures are more likely to seek information on policies previously adopted by legislatures with similar party leadership as well (Butler et al. 2017). High-speed internet has meant that over time states can more easily access a variety of sources and ideas, looking up model legislation from advocacy groups like the American Legislative Exchange Council or perusing what other states have enacted on the National Conference of State Legislatures website. More policy alternatives can be accessed more quickly, with more detailed information than in the past.

Broadband may also influence what policy scholars have called the public or systemic agenda (Kingdon 1995; Cobb and Elder 1972) through the availability of information and opportunities to communicate ideas and preferences with other citizens and/or policymakers. The internet has reshaped the media context, adding to the range of sources for citizens too. While local television and newspaper coverage highlight local candidates and issues (Prior 2006; Snyder and Strömberg 2008), digital media emphasizes national issues and events (Trussler 2019). Furthermore, large national websites are prioritized by search engines, which further nationalizes information networks (Hindman 2008).

Broadband and digital media have changed the information environment in other ways to diversify sources beyond traditional media and incorporate new forms of citizen engagement on policy. Information consumption is more interactive—easily shared, liked, and commented on, facilitating the ability of issues to go viral. Broadband provides new platforms for public discussion and displays of public opinion, often amplified by the traditional press in a hybrid media environment (Chadwick 2017). In this more participatory atmosphere, more issues may have a chance to emerge on the public agenda and possibly legislative agendas. The costs of media diversification and fragmentation have been debated (Sunstein 2007), but one impact may be the introduction of more ideas and issues in the public sphere.

In a study that examines hundreds of policies over decades, we cannot test the exact processes by which broadband use may influence state agendas and innovation, though the literature on public policy suggests multiple possibilities. We are able, however, to explore whether the growth of broadband is one factor influencing this trend of greater policy innovation over time, and whether it has a role in changing patterns for the diffusion of information.

Measuring Information Access and Innovation

Rising from less than 5% of the population with a home broadband connection in 2000 to over 80% in 2017, the past two decades have transformed information access, as can be seen in Figures 5.1, 5.2, and 5.3. Figure 5.1 shows the range of broadband subscriptions by state. In 2000, no state had more than 10% of the population with broadband. Subscription rates quickly rose to between 25% and 50% of the population with broadband by 2005, and then to every state with over 60% by 2015. While access to the internet has rapidly spread across the entire country, there are still large differences

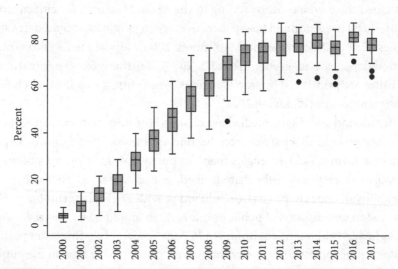

Figure 5.1 Percentage of adults with broadband subscriptions in US states, 2000–2017.

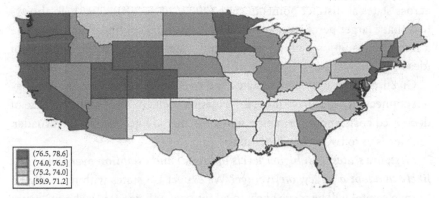

Figure 5.2 Change in broadband subscriptions by state, 2000–2017.

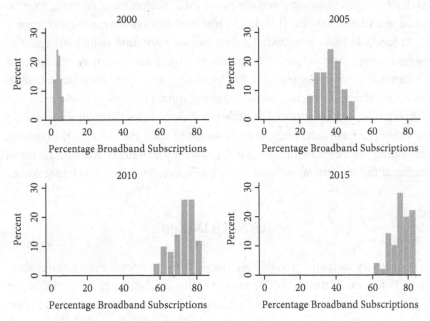

Figure 5.3 Histogram of broadband subscription rates in the states (2000, 2005, 2010, and 2015).

between the states, with gaps of as much as 25% between the states with the highest and lowest rates of broadband subscriptions.

As shown in Figure 5.2, states in some regions (West Coast, Mountain West, Midwest, New England) have experienced more growth in digital information access over the past 17 years. Figure 5.3 shows the distribution across states at distinct points in time (2000, 2005, 2010, 2015). To the extent that a larger percentage of a state's population is online, there should be a richer information environment within states to support the diffusion of ideas and innovation.

Given this variation across states and over time, states with greater digital connectivity may have more information-rich environments, because of decreased costs and information networks that are geographically broader. This leads us to two research questions.

First, are states with higher levels of broadband adoption over time more likely to adopt a policy, on average? We expect US states with more digital human capital will be more likely to adopt new policies across the spectrum of issues over time. As broadband connectivity becomes more prevalent, states have access to substantially more information from citizens, interest groups, and other states. This should result in states being more innovative.

Second, do more broadband subscriptions over time reduce the effect of neighboring states on policy innovation? Broadband use may disrupt past information flows. As the cost for legislators to learn about what distant states are doing decreases, the role of geographic proximity should decrease. Citizens also can be exposed to different state policies through social media or other platforms and become more aware of what states across the country are doing than ever before. Regional proximity will likely still be a factor in policy diffusion, but broadband could be expected to reduce its importance.

Research Design

The primary outcome variable we use is from the State Policy Innovation and Diffusion Dataset (SPID) collected by Boehmke et al. (2019) from policy adoptions over the past century. These data include information on thousands of policy adoptions for hundreds of policies, in all 50 states, across a variety of policy areas. The combined data for 2000–2016 include over 1,600 adoptions of 105 policies, and nearly 29,000 total policy-state-year observations. The dependent variable is a binary measure of whether a state adopted a policy in a year. Building on previous research, our time series models control for a host of state political, demographic, and economic variables (see table 5.1a for descriptive statistics, figure 5.1a for the distribution of observations and adoptions over time, and figure 5.2a for the correlations between independent variables used in the models).

We pair this with data on broadband subscriptions in the states from 2000 to 2016 from the US Census Bureau (the American Community Survey and Current Population Survey) to measure our primary explanatory variable.

To measure trends before broadband was widely available, we compare the current time period (2000–2016) to the last two decades of the 20th century (1980–1999) as a control. The 1980–1999 dataset includes more than 95,000 total policy-state-year observations. We also create a pooled model from 1980 to 2016, where broadband subscriptions are coded 0 before the year 2000. The combined dataset has nearly 170,000 observations and spans three and a half decades.

We use a pooled event history analysis to estimate the probability of a state adopting a policy. A state becomes at risk of adopting a policy after any state adopts a given policy (Berry and Berry 1990). If Hawaii adopts a policy, for example, the other 49 states become at risk of adopting that policy each year until they adopt the policy. Once a state adopts a policy, it drops out of the risk set. Each policy has its own risk set, and an event history analysis models the probability of a state adopting a policy in a given year. We follow Kreitzer and Boehmke's (2016) approach to pooling risk sets by including random intercepts for each policy. This method allows for pooling findings into a single, unified model while also recognizing that each policy has a distinct baseline probability of adoption.

Our key independent variable is a measure of the proportion of households with a broadband subscription in each state over time. Our first model includes this variable in a typical model of diffusion, with measures for previous contiguous state adoptions, as well as controls for legislative professionalism, wealth (state income per capita), state population size, diversity (percent Black in the population), educational attainment, a measure of unified party control of government, and state ideology (public opinion liberalism). These are all variables that have been commonly used in research on state policy diffusion and are among those that have most frequently demonstrated positive and significant effects on policy adoption, according to recent meta-analyses of hundreds of studies over decades (Maggetti and Gilardi 2016; Gilardi 2016; Mallinson 2021). To control for the rise of partisan polarization, we include a measure for the sum of the ideological distance between previously adopting states and a state at risk of adoption using Caughey and Warshaw's (2016) measure of policy liberalism. We use year fixed effects to account for variations by year and a cubic polynomial for the time that the state has been at risk of adopting a policy. Finally, we use the Desmarais et al.

(2015) measure of decayed latent diffusion ties to measure trends over time, which is an inferred network using previous policy adoptions that has been found to strongly predict policy adoption.

The first model shows the effect of digital information access (broadband subscriptions) on policy adoptions. The next set of models is the same specification, but with an interaction between broadband and previous geographically contiguous state adoptions. This tests whether broadband moderates the effect of proximity on policy innovation.

Instrumental Variable Models

Finally, an instrumental variable model is used as a robustness check to demonstrate that the relationship between the growth of digital information and policy diffusion is not spurious. We estimate a two-stage probit (instrumental model) that uses three geographic variables: state area, average elevation in a state, and Dobson and Campbell's (2014) measure of the percentage of a state's geography that is flat. When geographic barriers to building broadband infrastructure are high, we expect broadband subscription rates to be lower and vice versa. These three instrumental variables affect broadband subscription rates but do not affect state policy innovations (see chapter appendix for more discussion of the instrumental approach).

Simply put, people can only subscribe to high-speed internet access if the infrastructure exists. Geographically larger states may be more challenging for deployment, with sparsely populated rural areas that are more expensive to serve. Additionally, rugged terrain such as mountains will increase the cost of laying cable or fiber, potentially affecting provision or costs to consumers. The second stage of the model estimates policy adoption in the states and uses fixed effects for year, a cubic polynomial for duration, and fixed effects for policy to account for temporal dependence and different baseline probabilities of adoption between policies.

Broadband Adoption and the Growth of Policy Innovation

The results in Table 5.1 (model 1) show that both geographic proximity and latent network policy adoptions predict policy innovations over time and across the 50 states, consistent with previous research. States are more likely

Table 5.1 Pooled Event History Analysis of State Policy Innovations (2000–2016)

	(1) 2000–2016	(2) 1980–1999	(3) Pooled 1980–2016
Broadband Subscriptions	1.7326**		0.7365**
	(0.6056)		(0.3600)
Latent Ties	0.0829***	0.1003***	0.1315***
	(0.0220)	(0.0113)	(0.0085)
Geographic Contiguity	0.1585***	0.1878***	0.1766***
	(0.0273)	(0.0178)	(0.0118)
Ideological Distance	−0.0040	−0.0062	−0.0081**
	(0.0052)	(0.0040)	(0.0028)
Population	0.0538	0.1235***	0.0994***
	(0.0356)	(0.0262)	(0.0180)
Public Liberalism	0.0351	0.1516***	0.1532***
	(0.0367)	(0.0346)	(0.0205)
Unified Control	0.0469	−0.0824**	−0.0143
	(0.0552)	(0.0353)	(0.0253)
Income per Capita	−0.1043**	−0.0737*	−0.0677**
	(0.0424)	(0.0407)	(0.0240)
Legislative Professionalism	−0.0886*	−0.0955***	−0.1031***
	(0.0454)	(0.0288)	(0.0213)
Proportion Black	−0.0077	0.0088**	0.0080***
	(0.0156)	(0.0032)	(0.0024)
Percent High School	0.0116	0.0222***	0.0173***
	(0.0109)	(0.0048)	(0.0035)
Duration	−0.0227	−0.1954***	−0.0417**
	(0.0563)	(0.0305)	(0.0138)
Duration Squared	−0.0027	0.0292***	0.0016
	(0.0112)	(0.0045)	(0.0012)
Duration Cubed	0.0001	−0.0012***	−0.0000
	(0.0006)	(0.0002)	(0.0000)
Constant	−4.7559***	−3.6510***	−3.3370***
	(1.0185)	(0.4269)	(0.3147)
Var (Policy)	0.8403***	1.1815***	0.9534***
	(0.1328)	(0.1219)	(0.0794)
Observations	28,851	95,346	170,364

*$p < .05$, **$p < .01$, ***$p < .001$. Modeling includes fixed effect for year and random effects for policy. More policies began diffusing in the 1980s and 1990s than those that began diffusing in the 2000s, so the number of observations is larger for the earlier period. Many of the policies from the 2000s are still diffusing, while most from the 1980s and 1990s fully diffused. Broadband home adoption first measured by the US Census Bureau Current Population Survey (CPS) in 2000. Internet use dial-up first measured by the CPS in 1997.

to adopt a policy as the number of contiguous state adoptions increases, and as more states with latent ties adopt that policy. Both findings match prior research in policy diffusion. Ideological distance between a state and previous adopting states is either insignificant or negatively predicts policy adoption. The only significant controls are income per capita, legislative professionalism, and population size, all positively related to greater policy innovation.[3]

The key independent variable, digital human capital, measured by broadband subscriptions, increases the probability of a state adopting a policy. As the percentage of the population with broadband rises (across states and over time), so does the probability of a state being an early adopter of new policies. In fact, broadband subscriptions have a larger effect than contiguity or latent network ties (the variables have been standardized so that their effect size can be compared). These results support the argument that broadband adoption increases innovation in the US states.

As a comparison, model 2 (Table 5.1) replicates the analysis for the earlier time period (1980–1999) as a control case omitting broadband. Without including a measure of broadband use, the effect size for latent network ties and geographic proximity is similar. Model 3 pools data for the complete time period (1980–2016), confirming that broadband subscriptions remain a powerful predictor of policy adoptions, controlling for standard predictors of policy diffusion, including geographic contiguity, latent ties, population size, liberal public opinion, and more educated populations. The effect size remains large.

Figure 5.4 shows the predicted number of policy innovations varying broadband subscriptions in the population. The effect is substantively very large. The probability of a new policy adoption more than doubles from less than a 5% probability to over a 10% probability as a state's population rises from low to high levels of broadband subscriptions, all else equal. A roughly 10% increase in broadband subscription results in a 1% increase in the probability of policy innovation. An increase of 1% is substantively very large when considering that the baseline probability of adopting a policy in any given year is just under 5%. So, a 10% increase in broadband subscription rates increases the baseline probability of adopting any new policy by roughly 20%. Broadband subscriptions have a positive and substantively large relationship with new policy adoption.

The results in Table 5.2 show the interaction between contiguity and broadband subscriptions. The interaction between the two is negative and statistically significant across the three model specifications: (1) current time

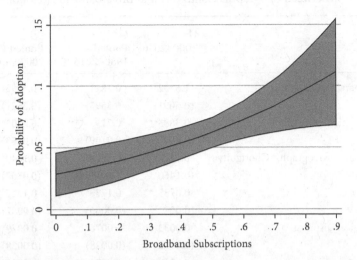

Figure 5.4 Predicted probability of adopting any given policy in a year (model 1, Table 5.1).

Note: Predicted probabilities are generated from the fixed component of model 1 (Table 5.1) and reflect the averaged random effects. Shaded area represents 95% confidence interval around the probability estimate.

period (column 1), (2) pooled time period (column 2), and (3) pooled time period with an additional binary variable (column 3). Model 3 adds a binary variable for years before 2000 as an additional robustness check to isolate variation between the two eras. This means that over time, increases in broadband subscriptions diminish the influence of geographic contiguity in the adoption of new policies.

Figure 5.5 shows the marginal effect of contiguity by broadband subscription rate (model 1, Table 5.2). The effect of contiguity is positive and significant; neighboring state adoptions increase the probability of policy adoption by roughly three-fourths of a percentage point with each additional contiguous state adoption. However, the marginal effect of this proximity decreases as broadband subscription rates surpass 60% of the population. Contiguity no longer positively predicts policy adoption when subscription rates are over 85%. Broadband has altered flows of information and reduced the role of geography in policy diffusion. The states with high levels of digital human capital appear to no longer rely heavily on neighboring states for policy examples and ideas.

Table 5.3 compares states with above-average and below-average broadband subscription rates for the current time period (2000–2016). We find

Table 5.2 Pooled Event History Analysis of State Policy Innovations with Interaction between Geographic Contiguity and Broadband Subscriptions

	(1) 2000–2016	(2) Pooled 1980–2016	(3) Pooled Pre- 2000 Dummy
Broadband Subscriptions	1.8702**	0.7211**	0.6893*
	(0.6091)	(0.3597)	(0.3621)
Geographic Contiguity	0.3885***	0.2199***	0.2205***
	(0.0999)	(0.0149)	(0.0153)
Broadband # Geographic Contiguity	−0.3478**	−0.1714***	−0.1649***
	(0.1460)	(0.0364)	(0.0367)
Latent Ties	0.0753***	0.1273***	0.1325***
	(0.0222)	(0.0085)	(0.0087)
Ideological Distance	−0.0033	−0.0078**	−0.0079**
	(0.0052)	(0.0028)	(0.0028)
Population	0.0545	0.0994***	0.0998***
	(0.0355)	(0.0180)	(0.0182)
Public Liberalism	0.0291	0.1515***	0.1474***
	(0.0368)	(0.0205)	(0.0208)
Unified Control	0.0482	−0.0153	−0.0100
	(0.0552)	(0.0254)	(0.0258)
Income per Capita	−0.1065**	−0.0673**	−0.0724**
	(0.0424)	(0.0240)	(0.0245)
Legislative Professionalism	−0.0865*	−0.1019***	−0.0998***
	(0.0453)	(0.0213)	(0.0216)
Proportion Black	−0.0084	0.0085***	0.0093***
	(0.0156)	(0.0024)	(0.0025)
Percent High School	0.0112	0.0182***	0.0197***
	(0.0109)	(0.0035)	(0.0035)
Duration	−0.0284	−0.0434**	−0.0500***
	(0.0565)	(0.0138)	(0.0141)
Duration Squared	−0.0007	0.0020	0.0026**
	(0.0112)	(0.0012)	(0.0013)
Duration Cubed	0.0000	−0.0000	−0.0000
	(0.0006)	(0.0000)	(0.0000)
Before 2000			−0.1596
			(0.4728)
Constant	−4.5298***	−3.3217***	−3.2798***
	(1.0255)	(0.3148)	(0.3543)
Var (Policy)	0.8495***	0.9608***	0.9449***
	(0.1341)	(0.0801)	(0.0793)
Observations	28,851	170,364	163,345

*$p < .05$, **$p < .01$, ***$p < .001$.. Modeling includes fixed effect for year and random effects for policy.

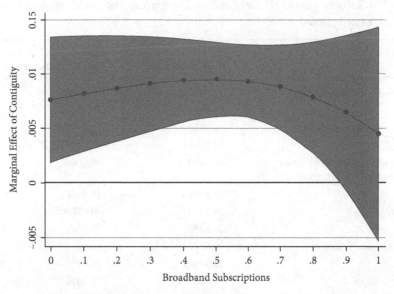

Figure 5.5 Marginal effect of contiguous adoptions by broadband subscription rate on policy adoption (model 1, Table 5.2).

that there are significant differences in what factors predict policy adoption in states with above-average and below-average broadband. The base term for contiguity is significant for both models, but broadband subscriptions only significantly predict policy adoption in states with above-average subscription rates. We find that broadband subscription rates reduce the effect of contiguity in high-broadband states but not in low-broadband states.

Table 5.3 also reveals differences in several other variables in the model comparing low- and high-broadband states. Latent ties and population size only predict policy adoption in states with below-average broadband subscriptions. Legislative professionalism lowers the probability of innovation in low-broadband states but has no effect in high-broadband states. Income decreases innovation in high-broadband states but not in below-average broadband states. Broadband adoption alters the political context of a state, which in turn has effects on the diffusion network. Additionally, we find more supporting evidence that broadband adoption reduces the role of geography in policy diffusion.

Finally, Table 5.4 shows the results for the two-staged probit model that treats broadband subscription rates as an endogenous predictor of policy adoption. In the first stage of the model, larger states have lower levels of

Table 5.3 Pooled Event History Analysis Comparing States with High and Low Levels of Broadband

	(1) High Broadband	(2) Low Broadband
Broadband Subscriptions	1.8465**	1.8893
	(0.8258)	(1.4229)
Geographic Contiguity	0.6872**	0.4120**
	(0.2301)	(0.1746)
Broadband Subscriptions # Geographic Contiguity	−0.7319**	−0.5983
	(0.3144)	(0.4546)
Latent Ties	0.0433	0.0789**
	(0.0284)	(0.0381)
Ideological Distance	0.0035	−0.0321**
	(0.0057)	(0.0131)
Population	−0.0113	0.2472***
	(0.0429)	(0.0729)
Public Liberalism	0.0149	0.1012
	(0.0459)	(0.0734)
Unified Control	0.0235	0.0998
	(0.0679)	(0.1013)
Income per Capita	−0.1344**	−0.0085
	(0.0488)	(0.1017)
Legislative Professionalism	−0.0382	−0.2292**
	(0.0523)	(0.0959)
Proportion Black	0.0892	−0.0101
	(0.4620)	(0.0164)
Percent High School	0.0173	0.0166
	(0.0296)	(0.0149)
Duration	−0.0637	0.2637
	(0.0688)	(0.1627)
Duration Squared	0.0021	−0.1183**
	(0.0135)	(0.0598)
Duration Cubed	0.0001	0.0108*
	(0.0007)	(0.0061)
Constant	−4.5859*	−5.0227***
	(2.6415)	(1.3348)
Var (Policy)	1.0815***	0.6631***
	(0.1821)	(0.1775)
Observations	18,972	9,879

$*p < .05, **p < .01, ***p < .001.$

Table 5.4 Two-Stage Probit (Instrumental Model) with Broadband Treated as Instrumental Variable

	Stage 1: Broadband Subscriptions	Stage 2: Policy Adoption
Broadband Subscriptions		4.3728***
		(1.0824)
Geographic Contiguity	−0.0002	0.0665***
	(0.0003)	(0.0147)
Latent Ties	−0.0003	0.0318**
	(0.0003)	(0.0120)
Ideological Distance	0.0003***	−0.0049*
	(0.0001)	(0.0029)
Population	0.0035***	0.0303*
	(0.0004)	(0.0180)
Public Liberalism	0.0146***	−0.0343
	(0.0003)	(0.0230)
Unified Control	0.0088***	−0.0046
	(0.0005)	(0.0299)
Income per Capita	0.0337***	−0.1684***
	(0.0004)	(0.0417)
Legislative Professionalism	−0.0044***	−0.0260
	(0.0004)	(0.0232)
Percent High School	0.0021***	−0.0005
	(0.0001)	(0.0064)
Duration	−0.0010	0.0000
	(0.0007)	(0.0357)
Duration Squared	0.0001	−0.0033
	(0.0001)	(0.0054)
Duration Cubed	−0.0000	0.0001
	(0.0000)	(0.0003)
Geographic Area	−0.0163***	
	(0.0009)	
Flatness	0.0005***	
	(0.0000)	
Average Elevation	0.0000***	
	(0.0000)	
Constant	−0.0163***	−1.9541***
	(0.0009)	(0.5810)
Observations		27,689
Wald test of exogeneity: χ^2	11.91 ($p < .001$)	

* $p < .05$, ** $p < .01$, *** $p < .001$. Fixed effects for year and policy included.

broadband subscription rates, and flatter states on average have higher rates of broadband subscription. States with an average higher elevation have somewhat higher rates of broadband subscription. The other demographic variables behave as expected. Wealthy, well-educated, and populous states have higher levels of broadband adoption.

After stripping out these factors from our measure of broadband subscriptions, the second-stage models predict state policy adoption. The percentage of households with broadband subscriptions is a large, positive, and statistically significant predictor of policy adoption in the states. The predicted probability of policy adoption ranges from 4 percent for states with low levels of broadband subscriptions (below 20%) up to 7.6 percent for states with near-universal levels of broadband (over 90%). The 3-percentage-point increases in the probability of new policy adoption is substantively large when considering that the overall baseline probability of adoption is very low. Many of the other variables in the dataset behave as expected. Adoptions from contiguous states or states with latent ties increase the probability of adoption, and more populous states tend to be more innovative. Income per capita predicts higher levels of broadband subscription but not policy adoption, which provides further support for our argument that the relationship between broadband subscriptions and policy adoption is not spurious. Through a variety of specifications, we have shown that broadband subscriptions strongly and positively predict state policy innovation.

Information and Policy Innovation in the Digital Era

Since Jack Walker's (1969) seminal study, diffusion scholars have cited the central role of information in the spread of policies between states. In recent years, some researchers have noted that more new policies are being adopted (Boehmke and Skinner 2012) and at a more rapid pace, and that information and communication technologies may explain these trends because they increase the speed at which information travels (Boushey 2010, 23). With longitudinal data on broadband subscriptions, we are able to examine directly the effects of broadband use and find compelling results that broadband is a resource that contributes to policy innovation and is changing patterns of information exchange across the states.

We demonstrate that the effect of information gathered through digital human capital is substantively large, with a 10% rise in subscription rates increasing the baseline probability of adopting a policy in a given year by roughly 20%. This relationship holds under other specifications including an instrumental variable approach that controls for factors that may cause both policy innovation and broadband subscriptions.

Results show that digital human capital not only accelerates the spread of new policies across the states but also alters the flow of policy information. States with high rates of broadband subscriptions can easily look beyond their immediate neighbors for policy solutions, and citizens and interest groups are increasingly able to operate at a national level. States with high levels of digital human capital are less reliant on contiguity for policy or other traditional sources of information, such as latent networks or legislative professionalism. Others have noted that proximity has played a less prominent role in the diffusion of policy innovations over the past few decades (Mallinson 2021; Desmarais et al. 2015; Garrett and Jansa 2015; Gandara et al. 2017), and broadband may be one cause of this apparent shift.

All of this suggests richer state information networks over time and a greater diversity of ideas. States that are more innovative, embracing more policies at an earlier juncture, may or may not engage in learning or create the most effective policies. But clearly digital technologies are transforming the processes of diffusion and innovation in the states with the most inclusive broadband use. Innovation at the state level shapes the environment for local actions and outcomes, including prosperity and growth. But the analysis here indicates that digital human capital at the state level can influence policies and outcomes across a range of issues. In the next and final chapter, we will consider what broadband's influence means for the future of communities and public policy.

Chapter 5 Appendix

Table 5.1a Summary Statistics for Diffusion Models

	Mean	SD	Min	Max
Policy Adoption	0.0448808	0.207043	0	1
Broadband Subscriptions	0.170132	0.2621549	0	0.8733091
Latent Ties	0.0053934	1.00635	−0.3658659	14.62182
Geographic Contiguity	−0.0033685	0.9961798	−0.5999165	7.068464
Population	−1.74e-09	1	−0.8569154	5.734591
Public Liberalism	−2.55e-10	1	−2.255907	2.489642
Unified Control	0.4441863	0.4968765	0	1
Income per Capita	−4.75e-11	1	−2.416561	3.826919
Legislative Professionalism	−0.0006806	0.9997199	−1.324718	3.834485
Proportion Black	6.296458	8.773737	0	36.3
Percent High School	69.36862	27.91463	0.7970402	92.8
Duration	7.070469	6.312314	0	33
Duration Squared	89.83661	140.8769	0	1089
Duration Cubed	1466.794	3359.106	0	35937
Observations	167,889			

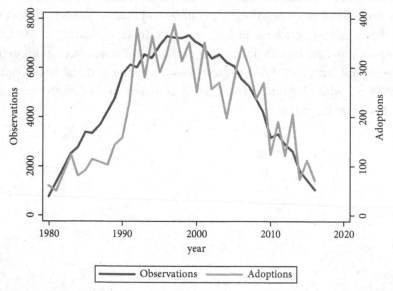

Figure 5.1a Adoptions and observations per year.

	(1) Policy Adoption	Broadband Subscriptions	Latent Ties	Geographic Contiguity	Population	Public Liberalism	Unified Control	Income Per Capita	Legislative Professionalism	Prop Black	Prop Latino	Percent High School
Policy Adoption	1.00											
Broadband Subscriptions	0.02***	1.00										
Latent Ties	0.09***	-0.14***	1.00									
Geographic Contiguity	0.10***	0.19***	0.13***	1.00								
Population	-0.00	0.05***	0.01	0.00	1.00							
Public Liberalism	-0.00	0.16***	-0.01	-0.09***	0.19***	1.00						
Unified Control	0.01*	0.18***	-0.06***	0.04***	0.08***	-0.13***	1.00					
Income Per Capita	-0.01	0.29***	-0.01	-0.05***	0.17***	0.59***	-0.07***	1.00				
Legislative Professionalism	-0.01	0.09***	-0.00	-0.01*	0.72***	0.48***	-0.03***	0.30***	1.00			
Prop Black	-0.01	-0.32***	-0.03***	-0.08***	0.03***	-0.06***	-0.01	-0.07***	-0.03***	1.00		
Prop Latino	0.02**	-0.46***	0.09***	-0.07***	0.22***	0.09***	-0.07***	0.02***	0.13***	0.08***	1.00	
Percent High School	-0.01*	-0.84***	0.14***	-0.18***	-0.05***	-0.00	-0.18***	-0.07***	-0.04***	0.17***	0.55***	1.00
Observations	28851											

Figure 5.2a Correlation between independent variables.
t statistics in parentheses. $^* p < 0.05$, $^{**} p < 0.01$, $^{***} p < 0.001$.

Checklist for Instrumental Variable Analysis

We follow the checklist developed by Sovey and Green (2011) when constructing our instrumental variable model.

1. Estimand: We argue that broadband's effect on policy innovation should be homogenous across states. We believe that broadband is the endogenous variable that causes more policy innovation, and that state geographic area, average elevation, and the flatness of an area affect the ease of installing broadband infrastructure but do not have a direct effect on policy innovation.

2. Independence: The next step is to establish "that an instrumental variable be correlated with an endogenous variable X, but not causally related to Y." We do not expect state geographic area, elevation, or flatness to be a cause of policy innovation in the policies in our dataset. When looking at descriptive data from the SPID dataset, the most innovative states in the country are California (3rd largest by area, 24th flattest), Minnesota (14th largest by area, 5th flattest), and Connecticut (48th largest, 44th flattest). Neither geographic area, flatness, nor average elevation predicts innovativeness in an event history analysis.

   However, we do expect geographic area and elevation change to affect broadband. Recent estimates from the Federal Communications Commission (FCC) show that the average cost is $27,000 per mile to install fiber lines to connect communities to the internet. So, on average, it will cost more money to reach the entire state population in geographically large states than geographically small ones. Installation is also cheaper over flat areas with less treacherous terrain such as mountains or hills that can make it more difficult to install fiber-optic cables.

3. We next use a series of diagnostics to evaluate the model. The first is the Wald test of exogeneity. The null hypothesis is that the endogenous variable (broadband subscriptions) is exogenous. The chi-squared statistic is 11.91 (p value of less than .001), meaning we can reject the null of an endogenous predictor. We next look to the f test for the first stage in the model. The null hypotheses is that the instrument is weak. With an f statistic of 6335.68, we can reject the null hypothesis of a weak instrument. The adjusted r-squared for the first stage of the model is also very high (.9703), meaning that we have created a powerful predictor of the endogenous variable.

4. Exclusion restriction: We argue that state size, flatness, and elevation do not have a direct effect on policy innovation. State boundaries were decided well before broadband infrastructure began spreading, so we can eliminate the possibility of broadband internet affecting state geography.

6

Choosing the Future

Digital Human Capital and Inclusive Innovation

Broadband is more than a technology—it's a platform for opportunity. No matter who you are or where you live in this country, you need access to advanced communications to have a fair shot at 21st century success.

—Jessica Rosenworcel (Federal Communications Commission 2019, 325)

In an age of rapid digital transformation, communities and society confront choices that will resonate far into the future. Technology has spurred local and national economic growth through innovation and the geographic concentration of human capital and digital skills; at the same time, it has led to rising inequality across places. While a community's human capital is typically measured in terms of college-educated workers, we have argued that there are multiple dimensions to human capital today, including the ability to access and use information technology (IT). This book examines the impact of internet use and digital skills as a form of human capital.

Broadband subscriptions, which reflect technology use rather than infrastructure, are a collective measure of this digital human capital. They are a measure of the digital capacity of communities, defined in this project as neighborhoods, cities, metros, counties, and states—a measure of the extent to which digital skills and information are widespread in a community. The future of communities need not be determined by rising inequality in the face of technological change. *Places that act to increase their digital human capital are choosing a path of inclusive innovation and opportunity.*

This book presents the first conclusive evidence that broadband adoption in the population is linked to economic growth and prosperity in counties and metros, whether urban, suburban, or rural. Public policy, including the

Choosing the Future. Karen Mossberger, Caroline J. Tolbert, and Scott J. LaCombe, Oxford University Press. © Oxford University Press 2021. DOI: 10.1093/oso/9780197585757.003.0006

National Broadband Plan (Federal Communications Commission [FCC] 2010), has been premised on the expectation that broadband use fosters economic opportunity for communities. The quote from Commissioner Rosenworcel at the beginning of this chapter demonstrates the significance of broadband for public policy, beyond a utility to be regulated. Until now, however, evidence for community-level benefits of broadband adoption has been limited because of a scarcity of reliable data over time. Measuring broadband subscriptions rather than deployment is critical because of what has been called the "subscription gap" (Tomer and Shivaram 2017)—the difference between broadband availability and the reality of those who can afford it and have the ability to use it. In this chapter we review the findings throughout the book and discuss place-based barriers in communities and their implications for public policy.

Drawing on nearly two decades of new data on broadband subscriptions, we have traced the digital past of the nation's states, counties, largest metros, and even neighborhoods to better understand policy choices for the future. Examining change over the last two decades reveals the degree to which broadband subscriptions may be causally linked to the impacts we uncover. Through time series analysis, we identify the extent to which increases in broadband adoption help explain subsequent changes in economic prosperity, growth, income, or employment. The evidence is compelling, as broadband's impacts are replicated across all types of communities and for different geographies. With time series data, we can also portray trends such as the flattening out of adoption rates in recent years, or stronger interaction effects for the presence of IT employment in metros in earlier years.

The outcomes we uncover here also contribute to our understanding of rising inequality across individuals and places in recent decades, laid bare during the pandemic. This inequality has concerned scholars across disciplines because of diminished economic opportunity and its effects on political participation and democratic institutions, but current explanations have focused on different aspects of the problem. These include federal tax policy (Bartels 2017; Saez and Zucman 2014); state economic policies (Franko and Witko 2017); human capital, decline of manufacturing, and technological change (Moretti 2012); economic and racial segregation (Florida 2017; Galster 2017; Kneebone and Holmes 2016; Owens 2016; Bergman et al. 2019); schools and family structure in a community (Chetty and Hendren 2017); urban zoning and housing policies (Trounstine 2018); and institutionalized racism at different scales (Michener 2018 on county Medicaid programs; Hero 2000 on states and demographic change).

None of the previous work on inequality has considered the role of the "digital divide"—inequality in digital skills and information—in spatially patterned disparities. Our theoretical justification builds on the concept of human capital, but it is broader than college-educated populations. Our models have included many previously identified aspects of the demography of place-based inequality, along with the structure of occupations in the local economy. And our research demonstrates that above and beyond these other factors, broadband subscriptions in the population matter for economic opportunity. They also matter for unemployment rates and community resilience in the wake of the pandemic, as we show toward the end of this chapter.

Broadband is part of the place-based narrative of innovation as well as inequality, exemplified by smart cities and burgeoning tech hubs. What has not been recognized, however, is that spatial disparities in *broadband adoption— and associated digital human capital*—may be a factor driving inequality of opportunity across places. The role of broadband use in the population has been unrecognized because it was unmeasured at the local scale throughout the course of the internet's over 30-year history.

Digital Human Capital for Innovation across Time and Place

The key to digital human capital is the ability to access and use information. First, that includes digital skills for jobs and entrepreneurship, for problem-solving and creativity. Second, the ability to use technology also unlocks access to information that can help individuals to increase other aspects of their own human capital, through education and training, networking, job search, better health, improved mobility around communities, and more. Third, digital human capital in turn supports networked information, producing spillovers and multipliers in the community.

Evidence of how broadband subscriptions matter for innovation is consistent across these chapters, utilizing rigorous methods—such as time series analysis—for urban, rural, and suburban communities, and for counties, metros, and states. In contrast with some past findings on broadband infrastructure (Kolko 2012), broadband subscriptions are significantly related to multiple benefits for residents as well as businesses. The prosperity index used in Chapter 2 on counties considers poverty levels, labor force participation, and job growth and income for the community as well as increases in the number of business establishments (Economic Innovation Group 2018). For metros in Chapter 3, we examine changes in prosperity defined as wages, productivity, and standard of living in the community, and growth as changes in gross metropolitan product, jobs, and employment in new firms (Brookings Metro Monitor, Shearer et al. 2018). The metro prosperity measures reflect benefits for residents and are strongly related to broadband adoption. Additional results for full-time employment for the metro analysis and median income for counties leave little doubt that broadband adoption affects opportunity in communities.

Counties

Chapter 2 offers a comprehensive view of internet use across all counties, in different contexts. Taking advantage of 2017 census data that includes all counties in the United States for the first time, Chapter 2 investigates the influence of broadband use in sparsely populated rural counties as well urban core and suburban counties.

- The analyses demonstrate that broadband subscriptions are positively related to the prosperity index, median income, and change in the prosperity index, controlling for other factors.
- Broadband use contributes substantially to prosperity; holding all else constant, moving from 20% to 80% of the population with a broadband subscription produces a 40-point difference on the prosperity index (on a scale of 0 to 100).
- A 1% increase in broadband subscriptions results in a $40 increase in annual median income across counties (average increase from 2016 to 2017 is $1,780), holding all other factors constant.
- Further exploration of the 2017 data shows that broadband subscriptions are significantly associated with higher median income for all types of counties: rural, urban, and suburban.

Some findings in Chapter 2 are based on cross-sectional analysis, given that new census data is available only for one year. We increase confidence in the causal influence of broadband in counties with additional tests.

- The Economic Innovation Group provides a measure of change in prosperity scores from 2007 to 2016, which they characterize as recovery from the recession. Broadband is a statistically significant predictor of positive *change* in the prosperity score as well. Counties with at least 70% of the population with broadband subscriptions recovered from the recession, while those below this threshold did not.
- To account for factors that might drive both economic outcomes and broadband connectivity, we employ two-stage causal models (instrumental variable regression) that use rural population and small business density in the first stage as instrumental variables. After stripping these factors out of our measure of broadband subscriptions, broadband adoption remains a significant predictor for community prosperity and median income.

We conduct time series analyses for counties in Chapter 2. Time series data allow us to observe change in economic outcomes (lagged by one year) following increases in broadband. This provides a stronger basis for concluding that broadband is statistically associated with the observed results.

- One model includes the 324 largest counties with data available from 2007 to 2012, and the second pools over 5,000 observations in counties over a 10-year period. This includes more than 800 counties with data available after the 2013 American Community Survey (ACS) as well as all counties in the 2017 data. In both models, broadband subscriptions significantly predict median income, with a 1-percentage-point increase in broadband adoption leading to an average $230 increase in median income in the pooled analysis.
- We interact two forms of human capital over time to predict income— the percentage of the college-educated population and broadband subscriptions. Increases in median income are greater in counties with more educated populations as well, indicating that both types of human capital affect economic outcomes.
- Finally, we replicate prior research comparing broadband subscriptions and availability of broadband, but for all counties rather than the non-metropolitan counties in Whitacre et al. (2014a). Consistent with that study, we find that broadband subscriptions are better predictors of economic outcomes than the number of broadband providers. Moreover, our more precise measures of broadband adoption demonstrate a

positive impact on median income at lower levels of adoption than the categorical FCC measures, and stronger relationships with median income overall. The categorical measures to some extent have obscured the strength of the association between broadband subscriptions and economic outcomes.

Metros

Metropolitan regions are the lifeblood of the national economy, and while some metros boast robust economic environments, others have languished, or are struggling to regain their footing after years of disinvestment. With broadband subscription data for the 50 largest metros that covers the entire 17-year period, Chapter 3 provides strong support for the power of broadband to shape local opportunity.

- Results show broadband use as a driver of increases in community prosperity and full-time employment in all instances, and growth under some conditions. Two different types of time series model specifications are used. One includes just a lagged term for broadband subscriptions in the metro in the previous year. The second includes the lagged term for broadband subscriptions and subscriptions in the current year. When we include both the lagged term and the current-year broadband subscriptions, the latter becomes a measure of "change in broadband use." Findings show that both broadband subscriptions in the previous year plus the change in broadband use result in positive change in economic outcomes for metros over time.
- Broadband subscriptions are also significant in predicting economic growth and full-time employment, but the results are less consistent for growth. Inclusive innovation measured by broadband use in the population is the strongest predictor of economic outcomes related to prosperity for residents.
- On average a 10% increase in broadband subscriptions each year increased the annual change in economic prosperity (productivity, wages, and standard of living) by 2 to 3 percentage points, controlling for other factors. A similar substantive impact was found for the annual change in broadband subscriptions. Similarly, a 10% increase in

broadband subscriptions resulted in a 2- to 3-percentage-point rise in full-time employment (change over time), all else equal.

Beyond this, we interact broadband subscriptions in metros over time with factors that have been identified in prior studies as significant predictors of economic growth—the share of IT employment and millennials in the population. These alternative measures of human capital do interact with broadband subscriptions and create threshold effects.

- Broadband subscriptions increase prosperity more in communities with significant younger populations, at least 30% aged 25 to 34. High levels of broadband adoption increase growth even when the percentage of millennials is low, but the combination raises the bar significantly.
- Interactions between broadband subscriptions and IT employment show that changes in prosperity are modest until there is at least 3% of the population in IT occupations. Once a metropolitan area reaches this 3% threshold, broadband use (and change in use) is strongly associated with increased prosperity. While broadband still has a positive effect on prosperity for metro areas with almost no technology jobs, the impact is much smaller.
- The interactive effects of broadband adoption and IT employment for growth differ over time. In the early to mid-2000s, moving from 12% of a metro's population online to 32% shifts this entrepreneurial growth by 5 points on the index for metros with 6% of the population working in IT. The same 20-point increase in the population online for nontech cities resulted in just a 1-point change. Yet this advantage for metros with high IT employment has narrowed in recent years.

Cities

Chapter 4 on cities and neighborhoods illustrates the inequalities in broadband adoption that exist across and within the nation's largest cities, at the forefront of gigabit networks and smart city advances. New neighborhood maps within cities show wide disparities in broadband subscriptions between affluent and poor zip codes, with up to 50-percentage-point differences in connectivity within cities. People in poor communities are much more likely to rely on an internet connection only on their cell phone, while residents of more affluent communities have both home broadband and smartphones.

- Fixed broadband subscriptions range from 85% of households in Seattle to only 48% in Detroit. Even when smartphones are included, one-third of households in Detroit and Miami have no personal broadband access at all.
- Racial and ethnic disparities are a feature of all cities, though the gaps vary. Connectivity, including cell phones, is only 50% to 70% for African American and Latino households in Detroit, Miami, Memphis, Milwaukee, Houston, and Dallas. In the six cities where at least 90% of the population has broadband (including cell phones), African Americans and Latinos fare better. With the exception of African Americans in Seattle (at 76%) and Latinos in Raleigh (at 78%), adoption rates for these two groups fall behind non-Hispanic whites but exceed 80% in highly connected cities.
- In Memphis there are neighborhoods where only 26% of households have fixed broadband (similar to rural Wheeler County, Georgia).
- Yet, there are neighborhoods with low rates of adoption even in the highest-ranked cities. In San Jose, some neighborhoods have rates of fixed broadband adoption of only 40%—nearly 50 percentage points lower than the city overall, and less than half the citywide average. Mobile-only internet use is also at 40% in some zip codes, demonstrating an effort by digitally disadvantaged residents to go online in some fashion.

Maps of broadband subscriptions (any type, including mobile) and smartphone-dependent internet use nearly mirror each other, demonstrating that cities are indeed developing new fault lines, reinforcing segregation and concentrated poverty. Higher rates of mobile-reliant internet use by low-income African Americans and Latinos have provided at least some internet access in these communities. Yet mobile use has not erased technology disparities because of data caps and the limited capabilities of cell phones for engaging in many activities online, including distance learning and résumé writing.

Chapter 4 includes an evaluation of neighborhood-level change in Chicago's Smart Communities program, using multilevel models and time series data (2008–2013). The methods for this evaluation continue the time

series models employed throughout the book. This research assesses evidence for neighborhood change, beyond program participants, to build a
culture of technology use, for local innovation, throughout the community.

- Controlling for demographic change over time, the nine neighborhood
  Smart Communities experienced statistically significant, higher rates of
  increase in internet use, home broadband, and online searches for jobs,
  health information, and mass transit information. Notably, the online
  activities with higher rates of increase, including health and mobility,
  contribute to human capital and economic opportunity.
- These differences are substantively large as well as statistically significant, between 9 and 11 percentage points higher.
- While the Smart Communities neighborhoods in 2017 still faced
  challenges for connectivity, the program demonstrated that change was
  possible at the neighborhood level, even for a short-term intervention.

States

Chapter 5 on states demonstrated that inclusive access to information is also
a source of policy innovation, long a topic of interest in state policy and federalism. State governments have been the focus of research on policy innovation and diffusion because of their potential for policy experimentation
and change in the American federal system. Information has been viewed
as a resource for the diffusion of innovation (Baumgartner and Jones 2005;
Boushey 2010) and has been measured indirectly in prior research by education in the population, legislative professionalization, or urbanization.
Broadband subscriptions are a more direct measure of digital information access for policymakers, interest groups, and the public, and may facilitate earlier adoption of new ideas and programs. We examine the effects
of broadband subscription growth for all 50 states over the 17-year period.
To measure the spread and adoption of new state policies, we use the new
State Policy Innovation and Diffusion Database (Boehmke et al. 2019) of
thousands of state policy adoptions from 1980 to 2016.

- Results show that as digital information access becomes more inclusive
  and a larger percentage of the population has a broadband subscription, the probability that a state government will innovate in public

policy increases, controlling for other known predictors. Widespread and democratic information access places states on the leading edge of policy innovation.

- Across hundreds of policies and thousands of policy adoptions, we show that growth in broadband subscriptions predicts early adoption of new programs. Our results help to explain the surge in the number of new policies and the faster spread of new ideas noted by other scholars in recent years.

- Broadband use alters the flow of information across states. Broadband's ability to connect distant communities has reduced the importance of geographic proximity in state policy diffusion. Legislators in Mississippi can quickly learn about policies in Alaska, for example, and interest groups can rapidly share policy information nationwide at a very low cost. While policies are still more likely to spread across neighboring states, broadband reduces the influence of proximity in policy innovation. Broadband use increases the diversity of sources for policy ideas.

Together these chapters demonstrate that inequalities in broadband use have consequences for both economic and policy innovation and outcomes. Broadband adoption supports innovation at different scales, from the state policy context to local economies. While technology firms and educated populations are indicators of innovative capacity in the digital economy, so too is the inclusiveness of broadband adoption and use in a community. Controlling for these factors, broadband subscriptions matter for prosperity, growth, and innovation. They magnify the effects of education, IT employment, and millennial populations in the most successful communities, but also offer benefits for places even in the absence of these other local advantages. With evidence over time and across communities, the findings here are more robust than possible in the prior literature on the deployment of broadband infrastructure.

Inclusive Broadband: Adoption Rates for Low-Income Households across Counties

Inclusive broadband entails getting people online regardless of income. Across the 3,100 counties, in the 2017 census data, the national average for households making more than $75,000 was 87% with a broadband

Table 6.1 Comparing Counties with Above- and Below-Average Low-Income Broadband Subscription Rates

Variable	Below Average	Above Average
Low-Income Broadband Subscription	35%	53%
Average Income	$42,300	$53,000
Average Prosperity Score	31.8	51.98
Percent Rural	58%	57%
Percent College Educated	16%	25%
% Black	11.6%	6.2%
% Hispanic	8.4%	9.3%
% Asian	.5%	1.9%

subscription. This drops to 69% for households earning between $20,000 and $75,000, and to just 44% for households earning less than $20,000 a year. Across counties, broadband subscriptions for this poorest group range from 0% to 80%. Counties where 8 out of 10 of the poorest residents have broadband subscriptions include affluent communities and places with many college students.

Table 6.1 compares averages across demographic and economic variables for counties that are above (right column) or below (left column) the mean for the variable measuring broadband subscription rates for households with annual incomes under $20,000. The average for counties below the mean is just 35% adoption, compared to 53% for the counties above the mean. While the percent rural and Hispanic is roughly the same for both under- and overperforming counties, places with higher African American populations have more poor people offline. But education stands out as the biggest difference. On average, 25% of the population has a college degree in counties with above-average broadband subscription rates for the poor, compared to just a 16% average for counties below the mean. That broadband and education continue to go together underscores our digital human capital argument.

Figure 6.1 presents a county map of the ratio of broadband subscriptions for households earning more than $75,000 annually to those earning less than $20,000 (see Table 6.2). Lighter-shaded areas have more inclusive broadband, where the differences in connectivity between the poor and

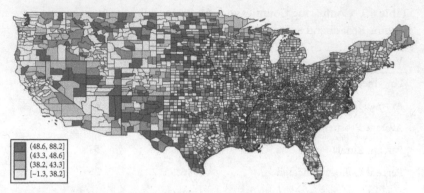

Figure 6.1 Differences in broadband subscriptions between high- and low-income households (lighter color indicates more equality).

Table 6.2 Inequality in Broadband Subscriptions by Income, Counties

Countries with Greatest and Least Inequality in Broadband Access by Income

10 Most Unequal Countries	10 Most Equal Countries
1. Loving, TX	1. Pitkin, CO
2. Mitchell,, TX	2. Nantucket, MA
3. Harding, NM	3. Lake of the Woods, MN
4. Faulk, SD	4. Issaquena, MS
5. Oliver, ND	5. Manassas Park
6. Wheeler, NE	6. Crockett, TX
7. Wibaux, MT	7. Fairfax, VA
8. Guadalupe, NM	8. Storey, NV
9. Kemper, MS	9. Logan, NE
10. Dimmit, TX	10. Teton, WY

more affluent income groups are small. Darker-shaded counties have stark differences in broadband adoption by income group. Some preliminary analysis suggests that counties with inclusive broadband, where even the poor are online, have the greatest economic growth. Future research is needed to explore these relationships.

Place Effects and Barriers for Digital Opportunity

What will it take to make all communities digitally inclusive? In this section we review different patterns of access and use across rural, Tribal, and urban communities to better define policy needs. There are place-based differences, of course, for barriers to broadband use in Wheeler County in rural Georgia and in urban Memphis neighborhoods. We review these place-based differences and opportunities first and then discuss more general needs for affordable and reliable connections and skills for digital human capital.

Availability, Reliability, and Speed

The FCC has estimated that over 21 million Americans do not have broadband speeds available in their areas, most of them rural (FCC 2019). It is well known that this FCC data understates infrastructure needs, especially in rural areas and on Tribal lands, as geographically large census tracts are counted as fully served even if a single person within the tract could get service (Government Accountability Office 2018). Using data collected from their software and Xbox, Microsoft estimated that nearly 163 million people do not use the internet at broadband speeds (Lohr 2018), especially in places that are rural or poor.

While rural communities are varied in their access to high-speed internet (Stenberg et al. 2009), lack of available connections is more common in rural than urban areas. Rural areas may lack private sector providers willing to undertake the expense of installing networks with few subscribers in sparsely populated areas with few economies of scale. Deployment costs increase in mountainous and other difficult landscapes, or where residents are distant from existing infrastructure (Salemink et al. 2017; Feld 2019). Availability is not the only challenge, however, as rural solutions such as satellite access can be more expensive and unreliable. If communities are served by a single internet service provider, there is a lack of competition over cost or quality of service. Rural residents generally pay higher prices for slower and less reliable service (Feld 2019; Pacheco and Ramachandran 2019). Mobile broadband can help to fill some gaps in connectivity (Prieger 2013). Yet cell phone service may be spotty in some rural areas or be financially out of reach for low-income residents (Strover et al. 2017). Rural libraries fill important needs for access in their communities but may be distant from residents of remote areas. Some rural libraries have programs to lend Wi-Fi hotspots, as there are often fewer nearby businesses that provide free Wi-Fi (Strover et al. 2017). While these can help to fill critical needs, as they surely did in parking lots and on school buses in the pandemic, hot spots still depend upon the reliability of cell phone signals (Strover et al. 2017). Speed and reliability of service also matter for what can be accomplished online, including for digital human capital in rural communities—for example, for distance education or telecommuting that requires streaming, or telehealth that demands even more bandwidth and precision.

The Rural Paradox

The US Department of Agriculture argues that better connectivity in rural areas can improve marketing, e-commerce, access to supply chains, labor productivity, information for business planning and management, and precision agriculture (US Department of Agriculture 2019). In small towns that act as hubs within urban regions, there is a potential to build "digital economy ecosystems" (Center on Rural Innovation n.d.). More generally, broadband availability and use can improve the quality of life in rural communities by bridging the distances that create what has been called a rural penalty in access to goods, services, markets, and employment. Telemedicine, distance learning, and telecommuting have great potential to fill unmet needs in these communities. Yet, the rural paradox is that some rural areas are among the places with the poorest connections and lowest rates of adoption (Salemink et al. 2017; LaRose et al. 2007, 2011).

The focus in the literature on rural broadband is often on the lack of availability, slower speeds, or the persistence of dial-up (Salemink et al. 2017), but that is an oversimplification of the policy challenge. Rural populations have a higher percentage of individuals who are low income, less educated, and older, all factors that have long been identified as related to digital inequality (Stenberg et al. 2009; Strover forthcoming). A 2015 study estimated that 38% of the gap between urban and rural broadband subscriptions was due to the lack of infrastructure, but a greater portion—52% of that gap—was attributable to the lack of adoption even when service was available (Whitacre et al. 2015). Even more than a decade ago, rural-urban differences in adoption were largely nonexistent between households of the same income level (Stenberg et al. 2009, 8); this same pattern exists today in the county maps of broadband adoption for households earning more than $75,000 annually (see Chapter 1).

Ignoring the issue of affordability and the need for training and support for less educated and older populations will perpetuate digital inequalities in rural areas. As the prior research by Whitacre and colleagues (2014a) suggested, broadband adoption—measured here as subscriptions—is a more meaningful gauge of place-based digital capacity than deployment alone, for rural as well as urban communities.

Deep Disparities in Tribal Nations

Tribal nations have some of the lowest adoption rates in the United States, as shown in the data for Apache County in Chapter 2. There has been a deficit of systematic and reliable evidence regarding the online experiences of residents in Tribal communities, including federal data that is limited (Government Accountability Office 2018; Duarte 2017; Howard and Morris 2019). One national survey of enrolled members with a primary residence on Tribal lands found that 38% of respondents went online primarily through smartphones, and that this was more common than use of other devices. Nevertheless, Tribal members used the internet when they could and engaged in a wide variety of activities online, including human capital–enhancing activities for jobs, education, and more (Howard and Morris 2019).

The 2017 census data on Tribal lands show for the first time that less than 27% of those living in the Navajo Nation have a broadband subscription and a home computer.[1] Overall, for the entire American Indian/Alaskan Native population of the United States residing on Tribal lands, only 53% had broadband subscriptions in 2017.[2] While the federal government has a trust relationship with Tribes that is embedded in federal policy (Howard and Morris 2019), the low rates of broadband adoption in Indian Country demonstrate a history of policy failure and neglect in honoring this commitment in broadband as well as other areas.

Yet broadband networks and digital human capital hold substantial promise for the future. In 2019, the FCC established a priority window for Tribal governments to apply for rights to unassigned spectrum over their lands—the 2.5 GHz band that had been traditionally reserved for educational services and was underutilized. Tribes with access to the 2.5 GHz band will control a resource for developing next-generation broadband, including 5G, in their communities (FCC 2018c; Brunston 2020).

Tribal control of broadband is an exercise of Tribal sovereignty and self-determination. Tribally owned enterprises, including broadband networks, provide jobs and support critical services for enrolled members. Broadband can also support other Tribal enterprises, which span a variety of industries and may be the largest employers in their counties, even for nonmembers of the Tribe (Atkinson and Nilles 2008). Some Tribal governments have already demonstrated leadership in this area by developing their own broadband networks, including Red Spectrum Communications of the Coeur D'Alene, Tribal Digital Village of the Southern California Tribal Chairmen's Association, Lakota Network of the Cheyenne River Sioux Tribe, and Gila River Indian Community (Duarte 2017, Chapter 4). These nations and others have also prioritized building digital human capital through affordable access, computer centers, job training, skills certification, and computer science programs in Tribal colleges (Duarte 2017, 116–117). Broadband is an important resource not only for economic development in Tribal communities but also for health care, education, governance, and language and cultural preservation, further supporting Tribal sovereignty (Morris et al. forthcoming).

Smartphones and the Less Connected in Urban Neighborhoods

There are three times as many urban households as rural households without broadband of any kind, including mobile and satellite services (Siefer and Callahan 2020). This became apparent in the pandemic, as urban as well as rural school districts rushed to provide devices and hotspots for unconnected students. Some urban communities lack adequate connections, though most have some form of broadband availability through cable, DSL, fiber, or mobile. Low-income urban areas may have lower-quality service with slower speeds, fewer providers, and less competition, as well as digital "redlining" of some poor areas (Holmes et al. 2016; Turner Lee 2019; Callahan 2019a). In most cases, however, urban disparities stem from lower rates of adoption in low-income neighborhoods because of cost or other barriers rather than an absence of infrastructure.

As Chapter 4 showed, reliance on smartphones as the primary form of internet access is common in low-income urban neighborhoods, including in cities that are otherwise highly connected. In focus groups of smartphone-dependent internet users conducted in Chicago in both English and Spanish, cost was the primary reason for not having fixed broadband at home for most respondents.[3] Participants spoke about limited use because of data caps and strategies to extend time online, such as regularly parking behind a sandwich shop late at night to use the Wi-Fi signal or looking for a chance to use a computer in an electronics store to avoid long waits at the public library. A survey of Chicago neighborhoods showed that, in fact, residents who lived in poor, racially segregated neighborhoods were more likely to participate in economic and civic activities with smartphones, but those who had fixed broadband subscriptions and home computers engaged in these human capital–enhancing activities even more (Mossberger et al. 2017). In their study of three Detroit neighborhoods, Fernandez and colleagues (2020, 18) similarly concluded that "While ingenious and routine use of smartphones was pervasive among those we interviewed and surveyed . . . we found that those who were highly mobile dependent were less likely to use the internet in ways that can counter socioeconomic divides."

Living in a high-poverty neighborhood reduces technology use across different types of communities, across racial and ethnic groups, and it has an independent effect over and above individual-level income, education, race/ethnicity, and age (Mossberger et al. 2006, 2013). Those who live in poor communities face a context that often lacks the support needed for acquiring digital human capital, with underfunded schools and libraries, and less access to jobs where they can develop their skills (Kaplan and Mossberger 2012). In cities and metropolitan areas, concentrated poverty has grown for both African Americans and Latinos since the last recession (Kneebone and Holmes 2016).

But cities, with their diverse populations and economies, have the potential to effect change as well. As McIlwain (2020) demonstrates in his history of African American technology innovators, the accomplishments of many who developed software, created content, and organized cybernetworks have gone unrecognized. Chicago, for example, has talented professionals and an opportunity to build the Black tech ecosystem discussed in Chapter 2 (Wilson et al. 2019), if choices are made that encourage access to institutions and resources in the city. Black organizations are already pioneering elements of such an ecosystem through accelerators and other programs, but broader support is needed too in education and the tech industry (Wilson et al. 2019). To fulfill the promise of digital human capital in these communities, inclusive broadband is a necessary foundation. Young people gathered in Flint's maker space who want to invent the future first need equal access to the platform on which it will be enacted.

Policy for Digital Human Capital in Communities

What policies are required to support digital human capital development? Residents need reliable and affordable broadband connections with adequate speeds and devices for performing activities online that promote human capital and opportunity. Equally necessary is the acquisition of skills for the use of devices and applications (including social media), for information literacy online, for managing privacy and security of data and interactions online, for creating content, and for decision-making and learning as technologies and skills evolve over time (Mossberger et al. 2003; Mossberger, Tolbert and McNeal 2008; Van Deursen and Van Dijk 2011; Helsper et al. 2015). In this section we review the state of recent policy and needs for the future of communities.

The Trump Administration and Rural Infrastructure

The emphasis in broadband policy at the federal level during the Trump administration was on rural deployment rather than broadband adoption. In January 2020, just before the pandemic, the FCC announced a $20 billion investment in rural broadband infrastructure programs over the next 10 years.[4] Funding for broadband appropriated by Congress prior to COVID-19 included $600 million in 2018 and $500 million in 2019 for rural and Tribal grants and loans in the US Department of Agriculture's e-connectivity pilot program. The pilot was advertised as having goals for increased productivity, improved operations, enhanced health care, educational opportunities, and competitive entrepreneurship in rural communities.[5]

Any assistance for communities represents progress, but these federal programs are inadequate to meet the needs for the economy and society in the 21st century. The focus on rural infrastructure alone neglects the problem of cost and skills in both rural and urban communities. Such policies reinforce institutional racism as well. The greatest number of households without broadband nationally are in metropolitan areas, and a majority in communities of color, whereas households affected by a lack of rural infrastructure are predominantly white (Siefer and Callahan 2020). Without attention to affordability and adoption as well, low-income communities of all backgrounds, in both urban and rural areas, will continue to lag behind.

The Subscription Gap and Affordability

Just as universal access to education has been important for economic development and democratic participation over the past century, universal access to the internet is critical for opportunity today. The most prevalent policy problem for achieving such access is the broadband subscription gap rather than the availability gap (Tomer and Shivaram 2017). Over 60% of households without any type of broadband have annual incomes of less than $35,000 (Siefer and Callahan 2020). Cost rather than lack of interest is a predominant reason for not having broadband in low-income communities, especially among people of color (Mossberger, Tolbert, Bowen, and Jimenez 2012; Goldberg 2019). Despite the increased need for internet access during the pandemic, 28% were worried about not being able to pay for home broadband, and 30% worried about paying for their cellphone. This rose to 36% and 39% for African Americans and 54% and 56% for Latinos (Vogels et al. 2020).

The affordability gap is spatially patterned and concentrated in low-income urban and rural communities (Tomer et al. 2017), affecting digital human capital in these places. Existing policies to address affordability, however, are a fragmented patchwork of modest and partial solutions that leave many uncovered and fall far short of the need for universal internet access. Solutions will inevitably require intergovernmental and intersectoral coordination on the part of multiple programs.

The federal program that most directly affects the affordability of broadband for low-income populations is the FCC Lifeline program, which has been a target of contention and attempted cutbacks in recent years (Siefer 2017). Through the Lifeline program, eligible low-income families can receive a monthly discount for either phone service (including smartphones) or broadband internet service. In most cases, households enrolled in the Lifeline program receive a $9.25 subsidy, though subscribers on Tribal lands can get an extra $25 discount per month (Gilroy 2017; Universal Service Administrative Company 2020). In 2017, only 28% of those estimated to be eligible for the program were enrolled (Durbin 2019). While there is room for outreach and expanded enrollment, the program only covers 13% of the average monthly advertised cost for broadband, and recipients are responsible for additional expenses (Chao and Park 2020).

Other programs that address affordability are nongovernmental: Comcast's Internet Essentials and a nonprofit program called EveryoneOn. In 2011, Comcast launched Internet Essentials, offering basic broadband of 15 Mbps at $9.95 per month along with training and a low-cost computer for $150. The program initially applied only to households with children eligible for free or reduced-price school lunches. Created as a concession to secure approval for a corporate merger, the program has expanded since then to cover more low-income households, including public housing residents, seniors, veterans, and recipients of many public assistance programs (Comcast n.d.; Wiggers 2019). In 2019, the program served 8 million customers in 2 million households and was the largest discounted broadband program in the nation, according to the company.

Since 2012, other internet service providers have offered discounts in conjunction with the national nonprofit EveryoneOn. Eligibility rules vary, with some accepting only students and others geared toward most social service recipients (EveryoneOn n.d.). Most of the providers have limited speeds, ranging between 10 and 18 Mbps, though the FCC's official definition of broadband is 25 Mbps download and 3 Mbps upload speed (EveryoneOn n.d.; FCC 2019).[6] Although EveryoneOn features multiple providers in 48 states, participating companies have collectively served 700,000 individuals as of 2019, totaling less than one-tenth of the Comcast program. The availability of discounts in a local area does not always mean that households in need can obtain them. Most companies, including Comcast, exclude households that have been customers within the past 90 days or that have unpaid bills.

During the COVID-19 crisis, the FCC asked broadband service providers to sign on to a pledge to not terminate service, to waive late fees, and to open Wi-Fi hot spots for public access during the pandemic (FCC 2020b). The pledge was originally set to expire at the end of April and then extended to June 30, 2020. Additionally, some companies offered free or discounted service for a limited time to those who had not purchased broadband before. But the offers varied by providers and regions of the country and often excluded people for many of the reasons previously mentioned. In other cases, individuals were charged or were denied the free or discounted offers (McCabe 2020). Over 2,200 complaints about the Keep America Connected Pledge were filed as of early May (Kelly 2020). As the pledge expired at the end of June, COVID-19 cases spiked upward in many states and preparations for the school year reverted to remote learning in many places.

Costs and Competition

Broadband in the United States is relatively expensive in comparison with other developed countries. There is little competition in the US broadband market, where only one or two providers generally serve an area, resulting in both slower and more expensive service. The Organisation for Economic Cooperation and Development (OECD) reported monthly costs for a "basket" of services with low data usage in 2017 of over $46 per month in the United States for fixed broadband, in contrast with an OECD average of less than $28 (OECD 2019). Similarly, the Open Technology Institute (OTI) compared broadband costs and quality in cities internationally and found that US cities were by far more expensive than their counterparts in other countries (Chao and Park 2020). Private sector plans in the United States also lack transparency, with complex pricing (such as bundling) and hidden costs and additional fees that make it difficult to compare pricing across providers. The OTI report compared 760 plans in 28 cities in North America, Europe, and Asia and found that European cities had the most affordable plans and Asian cities offered the most value, with faster speeds for the cost. While this is based on a small sample of cities, the results align with the older OECD data. Greater competition is needed in the United States to provide cheaper internet as well as bandwidths needed for smart city applications and innovative use (Crawford 2018, 209).

Some cities and counties have developed publicly owned community networks to fill gaps in availability, speed, and affordability. There are more than 500 publicly owned and 300 cooperatively owned networks (Community Networks 2020). While these tend to be in small cities, rural regions, or Tribal communities, Chattanooga, Tennessee, is an example of a larger city with a gigabit speed municipally owned network (Community Networks n.d.). In contrast, rural Alexander County, North Carolina, is developing a fixed wireless network in a mountainous area. The 14 US cities in the OTI's study of costs included several cities with municipal broadband, such as Ammon, Idaho; Chattanooga, Tennessee; Wilson, North Carolina; and Fort Collins, Colorado. Ammon, Idaho, was the least costly US city in the study, and the other municipal carriers were more affordable than other providers in their area for the speeds and services provided (Chao and Park 2020).

There are other models where cities collaborate with private sector providers to bring faster or cheaper service to their residents. These include various public-private partnership arrangements, sharing of local assets such as utilities poles or right of way to reduce costs, or working with community anchor institutions such as hospitals or schools to pool resources (Feld 2019). Far from encouraging competition, either public or private, 19 states have imposed either outright bans or restrictions on municipal networks (Feld 2019; Community Networks 2019). Incumbent service providers have lobbied hard in state legislatures to prevent greater competition (Crawford 2018, 174).

Coupled with rising income inequality in the United States, high-cost broadband limits the potential for human capital development and reinforces poverty of place. Increased competition and cheaper broadband for all are important parts of the solution, as well as discounts for poor households.

Digital Skills and the Future of Work

Beyond universal and affordable access, however, digital human capital requires the development of digital skills. This includes technical competence (operational and navigational skills), information literacy, communication skills, content creation, online collaboration, problem solving, and strategic use (Van Deursen and Van Dijk 2011; Laar et al. 2019; Schradie 2011; Helsper and Eynon 2013). Once individuals have gained access to the internet, a second-level "digital divide" in skills and uses of the internet (Hargittai 2002)[7] results in unequal outcomes or "returns" from internet use (Scheerder et al. 2017; Van Deursen and Helsper 2015). Internet skills and uses are related to other inequalities, including educational disparities (Van Deursen and Van Dijk 2011; Helsper and Eynon 2013) and other forms of social exclusion (Helsper 2012; Ragnedda 2019). This suggests the need for training and support for digital human capital, as well as broadband subscriptions. At a deeper level, though, it also points to the necessity of addressing fundamental educational disparities across neighborhoods and communities, which affect communities of color disproportionately because of segregation and educational funding that depends heavily on local resources (Orfield and Lee 2005; Orfield et al. 2004).

Moreover, rapidly changing workplaces and communities require lifelong learning opportunities for all to adapt and develop skills for the future. Trends toward automation, remote work, and online commerce have been greatly accelerated during the pandemic (Markle 2020; Frey et al. 2020; Smith 2020; Muro 2020; Lohr 2020), and they promise to deepen inequalities for less educated workers (Markle 2019, 2020). Harvard economist Lawrence Katz has posed the choice as whether "we going to take this moment to help low-wage workers move into the middle class and give them skills to thrive" (Lohr 2020).

Increasingly, access to jobs with benefits and a career path requires some level of digital skill (Shearer and Shah 2018; Horrigan 2018), and the occupations in most demand are jobs requiring these skills outside the technology industry (Smith 2020). "American jobs are undergoing massive technological transformation, with even entry-level workers now expected to use all manner of digital devices and equipment," according to a report from the National Skills Coalition (2020). "Examples include restaurant workers being trained in food safety using virtual reality goggles, home health aides using tablet computers to report patient information, retail clerks using smartphone apps to process returned items, and manufacturing workers using augmented reality to assemble parts." Yet, one-third of US workers have limited or no basic digital skills. Even younger workers in the United States lag behind their international counterparts in digital skills (Bergson-Shilcock 2020). Proposals for upgrading other types of job skills depend heavily on internet use for online degree programs, skills certification, vocational training, and coaching (Smith 2020; Markle 2020). As the Markle Foundation has argued, one part of the solution for equipping workers with needed skills must be addressing the digital divide (Markle 2020), including affordable broadband subscriptions and devices that provide resources for skill development.

Federal Policies for Supporting Digital Skills

Proposals have been advanced in Congress and outside government to provide federal funding for workforce development, including digital training (Lohr 2020; Markle 2020). Digital skills have been typically addressed at the local level, however, through schools, libraries, and community-based organizations. These programs depend heavily on local or nonprofit funding, and poor communities may be challenged to offer adequate training and support. There is currently no federal funding, however, to support skill development for schools or in the community. Some emergency funds in the CARES (Coronavirus Aid, Relief, and Economic Security) Act passed at the beginning of the pandemic assisted libraries (Benton Institute 2020) and a proposal passed by the House would fund training programs (Callahan 2020). Emergency programs are further discussed next.

Federal Emergency Assistance and Beyond

The pandemic may mark an inflection point for policies to address digital human capital. Though the programs are temporary, emergency efforts have advanced debates about more permanent solutions. At the time of this writing, for example, the Biden administration has proposed $100 billion for broadband as part of its infrastructure initiative.

Emergency funds already appropriated support broadband use in households, schools, libraries and health care institutions. The three COVID relief bills passed by Congress between March 2020 and March 2021 allocated just under $450 million for telehealth, $90 million to the US Department of Agriculture for rural telehealth and education, $285 million for broadband in historically Black, Hispanic-serving and Tribal colleges, $50 million for technology and support in libraries, $1 billion for Tribal broadband networks, and $400 million added to prior USDA funding for rural broadband deployment (Benton Institute 2020, 2021a). Schools and libraries were authorized to spend over $7 billion in E-rate funds on broadband infrastructure, connectivity and devices, including for home use by students and patrons (Benton Institute 2021c). Previously these funds only supported broadband service in schools or libraries.[8]

Most significant for addressing affordable access, however, is the $3.2 billion Emergency Broadband Benefit (EBB) administered by the FCC. The EBB provides up to $50 per month to subsidize broadband subscriptions ($75 per month in Tribal communities) along with subsidies for computers, laptops or tablets that are Wi-Fi-enabled and capable of video-conferencing. Eligible households include recipients for many different social service programs, those with incomes up to 135% of the poverty threshold, or those who experienced layoffs or furloughs during the prior year (Benton Institute 2021b). While the benefit is temporary and will terminate within 6 months of the end of the COVID emergency, the higher broadband benefits and other provisions are contained in the Moving Forward Act that has already passed the House. That legislation also includes $1.3 billion over 5 years to support digital skills and training in states and local communities. If enacted, the bill would promote competition by protecting local governments from state laws prohibiting municipal broadband and would require greater transparency in pricing on the part of providers (Callahan 2020).

Funding has been fragmented across programs during the pandemic, many of which are temporary. Scattered throughout the various programs and proposals, however, the potential for more comprehensive solutions emerges, including infrastructure, devices and training. Affordability is on the agenda with expanded subsidies for low-income households and proposals for greater competition and more transparent pricing. To make the most of this moment, a holistic approach to digital human capital will be needed, and much of this will require action by other levels of government as well as in the private and nonprofit sectors.

Collaboration for Community Capacity

Solutions for promoting widespread digital human capital must be intergovernmental and intersectoral if they are to succeed. Both states and local governments have been active in addressing broadband use through programs and partnerships, but with limited federal support and coordination in recent years, this has produced a patchwork of efforts with many missing pieces and incomplete coverage.

The National Telecommunications and Information Agency (NTIA) of the US Department of Commerce has organized a State Broadband Leaders Network (SBLN) to engage the states in broadband and digital inclusion. According to the NTIA, the "State Broadband Leaders Network is a powerful forum for connecting local government, industry and stakeholders across the country that are focused on broadband activities. . . . The SBLN also holds regular meetings for states to improve funding coordination, align policies, and address barriers to collaboration across states and agencies." (NTIA 2020b).

There is wide variation, however, in the extent to which states have dedicated staff or resources to these issues. Forty-seven states have legislation designating an agency with responsibility for broadband policy (Pew Charitable Trusts 2019).[9] Nearly three-quarters have a dedicated broadband office, and others have task forces or councils.[10] Half of the states have broadband plans (Stauffer et al. 2020), but not all of them have funded the plans or programs that have been enacted.

The Broadband Research Initiative of the Pew Charitable Trusts has identified nine states that engage in what it calls "promising practices":

- Stakeholder engagement
- Establishment of policy frameworks
- Planning and local capacity building
- Funding and operations
- Program evaluation

The featured states are California, Colorado, Maine, Minnesota, North Carolina, Tennessee, Virginia, West Virginia, and Wisconsin (Stauffer et al. 2020). An analysis of the impact of state programs found that they were significantly related to changes in broadband availability between 2012 and 2018. The presence of state-level funding and state offices was positively related to availability measures, and policies restricting municipal broadband had a negative effect (Whitacre and Gallardo 2020).

State programs tend to emphasize broadband deployment and rural communities (Pew Charitable Trusts 2019), and we argue that this is a good first step, but more work supporting adoption is needed as well. There are some exceptions, where states support programs that promote training, public access, and affordable solutions too. The California Public Utilities Commission, for example, has grants set aside specifically for broadband adoption (California Public Utilities Commission 2020) and collaborates with the nonprofit California Emerging Technology Fund (Stauffer et al. 2020). State responsibilities for workforce development, small business, education, and health also provide platforms for promoting broadband uses that contribute to human capital in communities.

Yet both state and federal governments have at times placed roadblocks in the way of local efforts. State pre-emption of municipal broadband is just one example. City governments have raised money from fees for right of way or cable franchises to provide grants or fund broadband adoption programs in their cities. These are precisely the types of fees restricted by the FCC for the rollout of 5G—a restriction challenged in the courts by San Francisco and Portland (Callahan 2019b).[11] Cities and counties need more autonomy to determine the policies that best respond to their needs and their own local context.

City and county governments have played an important role in promoting broadband adoption in libraries and schools. They have convened actors and resources across public and private sectors, drawing up strategic plans, forming local coalitions, and dedicating staff in some cities. The National Digital Inclusion Alliance (NDIA) is an organization involved in promoting digital skills and access, and the group has awarded the "Trailblazer" designation to local governments that have shown a commitment of their own resources, including staff and funding. The list (Table 6.3) demonstrates how some of the most active cities address this issue with their own resources.

As Table 6.3 shows, all Trailblazer cities offer some funding for digital inclusion programs, and the categories (not shown here) include funding for community groups, wireless networks, and other projects. Most cities are part of local coalitions that include public and private sector organizations. Local research includes surveys of residents on internet use or evaluation of programs. Almost all Trailblazers are major cities or counties, and they demonstrate local initiative on this issue.

Many of these cities and others described in Chapter 4 include places with high citywide rates of broadband adoption. These are cities where the community understands the critical need for broadband use. Some local coalitions have also received support from foundations like Rockefeller (New Orleans), Knight (Philadelphia and Detroit), and MacArthur (Chicago), as well as corporate donations in cities such as San Jose. While many big tech firms have corporate giving programs, there is a need for the technology industry to play a more strategic role in promoting broadband adoption and skills. Some recent programs have offered free or low-cost skills training, as Microsoft has done in its postpandemic partnership with LinkedIn and GitHub (Smith 2020). There are other examples. Grow with Google provides free training and resources for job seekers, businesses, start-ups, veterans, would-be coders, and teachers. The latter includes basic digital skills training for classrooms and augmented reality.[12] Empower by GoDaddy offers digital and entrepreneurial training and coaching for small businesses, focusing on underserved communities.[13] Microsoft's experimentation with the TV white space spectrum for rural connectivity is intended to offer affordable access that is adequate for cloud computing, artificial intelligence, and other contemporary applications.[14] Technology companies, which have benefited

Table 6.3 Digital Inclusion Trailblazer Cities, 2020

Local Government Name	Has Staff	Has Plan	Coalition	Survey Research	Has Funding	Affordability
+ City and Country of San Francisco	✓	✓			✓	✓
+ City of Austin	✓	✓		✓	✓	
+ City of Boston	✓	✓		✓	✓	
+ City of Chattanooga and Hamilton County	✓	✓	✓	✓	✓	
+ City of Long Beach	✓	✓	✓	✓	✓	
+ City of Portland	✓	✓	✓	✓	✓	
+ City of San Antonio		✓	✓	✓	✓	
+ City of Seattle	✓	✓		✓	✓	
+ District of Columbia	✓	✓	✓	✓	✓	
+ Louisville Metro Government	✓	✓	✓	✓	✓	
+ Provo City	✓	✓			✓	
+ Salt Lake City	✓	✓			✓	
+ City of Detroit	✓	✓	✓	✓	✓	✓
+ New York City	✓	✓			✓	✓

Source: National Digital Inclusion Alliance (2020).

from the digital revolution, have data, knowledge, and financial resources to promote more equitable participation in the face of accelerating social change. But government is responsible for ensuring the public good and needs to guide policy for the future.

Data and Research for Evidence-Based Policy

Along with the need to enact policies to support digital equity, it is essential to pursue research to inform policy decisions. This book has emphasized economic outcomes for digital human capital. As a general-purpose technology, however, broadband has many other uses and potential impacts for communities—for health, education, political participation, civic engagement, and more. The data we have estimated over time and used in this volume, as well as the ACS, provide better information than in the past for understanding the role of digital human capital for a variety of policy-relevant community outcomes.

Broadband Subscriptions and Other Policy Impacts

Broadband use has been called a meta-determinant of health (Schartman-Cycyk et al. 2019, Appendix D) because it affects all six social determinants of health: economic stability, education, transportation, food, housing, and social circumstances (Benda et al. 2020; Sheon and Carroll 2019). Broadband also plays a direct role in health care through telemedicine, online health information, health wearables and apps, patient portals, remote monitoring, remote diagnosis, and more (Strover forthcoming). Research using the FCC's categorical data on broadband subscriptions examined changes in self-reported health outcomes for 92 metropolitan and micropolitan areas from 2002 to 2009. The study revealed that broadband was associated with positive changes for 7 of the 24 measures (Whitacre and Brooks 2014). With more precise data on broadband subscriptions over time and across multiple geographies, there is the potential to explore broadband's effects across more communities and with greater confidence. As a public health determinant integrally linked to other aspects of the local context, more research is needed on broadband adoption, broadband uses, and community health outcomes.

Prior to the pandemic, concerns about the "homework gap" in student internet access at home was already a prominent issue for schools in poor communities (Auxier and Anderson 2020). Recent research has demonstrated how broadband access at home matters for students. One study of middle school and high school students in rural Michigan showed that controlling for other factors, those with broadband were more likely to perform more educational activities online at home, have higher grades and higher standardized test scores, and attend college (Hampton et al. 2020). This suggests that communities with more widespread and more equitable broadband use should have better educational outcomes, but the community-level evidence has been mixed. In part, this research has relied on measures of broadband availability, which says little about the ability of households to afford broadband. In their study of the introduction of new broadband infrastructure in North Carolina, Vigdor and Ladd (2010) concluded that broadband availability in communities had slightly negative impacts for reading and math. In contrast, Dettling and colleagues (2018) found that increased broadband availability at the zip code level increased SAT scores and students' admission exam scores. Prior research does not always show consistent results (Hampton et al. 2020), and with better place-based data on broadband subscriptions, we can gain a more informed view of how broadband affects the educational environment in schools and communities. These are just a few examples of how broadband adoption may matter for communities and their residents in two critical policy areas, health and education.

Data on Specific Uses of Broadband

Additional data on specific uses is also needed to explore the benefits of broadband. Neither the ACS nor the FCC data address specific activities online, such as use of the internet for health information, homework, and job search. One approach is local surveys, as discussed in the Smart Communities evaluation in Chapter 4, but these do not allow for comparisons across communities.

New forms of data derived from the internet might also inform our understanding of broadband use, such as search engine use, social media activity, and data shared by technology companies. Google searches have been used to identify public health concerns in cities or counties (Stephens-Davidowitz 2020; Philpott et al. 2020).[15] Machine learning has been used to understand spatial patterns for personal networks in Facebook, and this could be important for outcomes around civic engagement, voting, social capital, and more (McAuley and Leskovec 2012).[16] Government data on use of city websites or Open 311 service requests might be used to understand impacts of technology use (or the lack of it) on access to government services across neighborhoods.

As the internet became a lifeline for education, remote work, telehealth, and social interaction during the pandemic, data tracked by private firms provided a window into the role the internet played. During March 2020, Google searches for online food deliveries quadrupled worldwide, and online shopping doubled (Ungerer and Portugal 2020). In May, online spending was up 77% year to year in the United States according to Adobe's Digital Economy Index, accelerating trends in the growth of e-commerce by four to six years (Koetsier 2020). Social scientists used data from credit card companies and rideshare companies, mobility data from smartphones, and other indicators of digital activity to describe trends during the pandemic (Wellenius et al. 2020; Painter and Qiu 2020; Lyu and Wehby 2020; Chetty et al. 2020). Such nontraditional forms of "big data" might also help to describe the effects of broadband use in different communities in the future.

The data on domain name websites discussed in Chapter 2 demonstrates the potential for measuring outcomes for more specific types of broadband use. We turn to this data, asking whether counties with higher rates of broadband use had lower unemployment rates at the height of the shutdowns in April 2020.

We explore this question using both data on broadband subscriptions from the ACS and the density of domain name websites by county. Domain name websites offer more specific measures of economic activity online, including digital participation of microbusinesses, start-ups, and part-time entrepreneurs. Through collaboration with GoDaddy, the world's largest registrar of domain names, the researchers gained access to de-identified monthly data on over 20 million websites from November 2018 to September 2020, and we track the density of domain name websites by zip code and county as a measure of online commercial activity. GoDaddy administers approximately half of the domain name websites in the United States.

Comparisons with small business data from the census as well as surveys of domain name owners demonstrate that many of these websites represent microbusinesses currently not counted in government data (see Mossberger et al. 2020). We measure domain names that are actively being used (not "parked") and ask how (1) broadband subscriptions and (2) the density of domain name websites in a community (the number websites per 100 people in a county) are related to unemployment rates over time, controlling for other known factors.

The outcome variable is the Bureau of Labor Statistics' monthly unemployment rates for counties in April 2020. The statistical models include the same extensive controls as used in previous chapters of this book. These include variables measuring demographic factors (size of racial and ethnic groups, educational groups, age groups), percentage of the population employed in different economic sectors, and population.

To provide a multifaceted measure of broadband adoption and use in a community, we also include an interaction term for broadband subscriptions multiplied by the number of domain name websites per capita lagged on month.

As shown in Figure 6.2a and Table 6.4, counties with more broadband subscriptions, holding all else equal, have higher April 2020 unemployment rates. But adding just 1 commercial domain name website per 100 people leads to lower unemployment rates at the height of the pandemic. Counties with more domains per 100 people and more broadband connections had significantly decreased unemployment rates in April, across all income levels (see Figure 6.2b and Table 6.4, column 1). A similar result is found for US metros but not shown here.

Column 2 of Table 6.4 repeats the model for only US counties in the top 50% for COVID-19 cases as of April 2020. Column 3 of Table 6.4 includes a control for the percentage of low-income retail jobs lost. Retail was an industry that was hit especially hard, and low-wage workers suffered disproportionately (Parilla et al. 2020). The same results are found, with counties with higher broadband subscriptions and more digital commercial activity more resilient during the pandemic, with lower unemployment rates.

At 56% of the population with a broadband subscription, counties start to get beneficial effects from incremental increases in commercial websites on overall unemployment. Holding all else equal, adding 1 additional domain name site per 100 people results in a nearly 0.4% reduction in April unemployment when moving from counties with the lowest to highest broadband penetration. For instance, adding 5 sites per 100 people in a county with high broadband could reduce unemployment by 2 percentage points. Overall national unemployment rates in April 2020 were 14.37%, up from 3.5% in February 2020. Adding 1 domain name site per 100 people to Lake County, Ohio, with a broadband

(a)

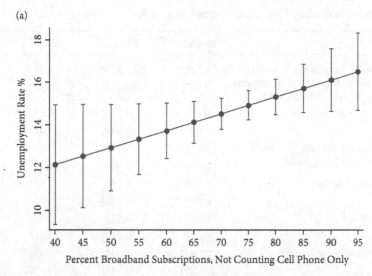

Figure 6.2a Predicted Unemployment Rate by Broadband Subscription Level

(b)

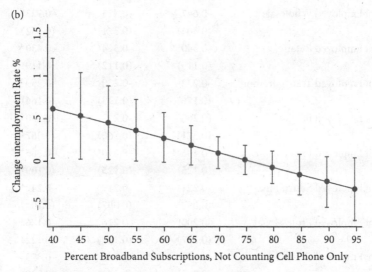

Figure 6.2b Marginal Effect of adding 1 Domain Name Website per 100 People at Different Broadband Subscription Levels

Table 6.4 Predicting April 2020 Unemployment Rates across US Counties

	Model 1	Model 2	Model 3
Venture Density (Domain Name Commercial Activity) Three-Month Rolling Avg (Jan–March 2020)	1.310* (0.603)	1.313* (0.610)	1.423* (0.602)
Percent Broadband Subscriptions	0.169*** (0.042)	0.156*** (0.042)	0.123** (0.044)
Interaction of Venture Density × Broadband Subscriptions	−0.017* (0.008)	−0.017* (0.008)	−0.019* (0.008)
Small Business Density	0.129 (0.082)	0.133 (0.081)	0.151 (0.081)
Percent Black	0.034 (0.021)	0.023 (0.019)	0.021 (0.019)
Percent Native American	0.063* (0.028)	0.054* (0.026)	0.056* (0.025)
Percent Asian	0.005 (0.082)	0.010 (0.082)	0.001 (0.078)
Percent Hispanic	0.089*** (0.025)	0.087*** (0.025)	0.115*** (0.029)
Percent Employed Agriculture	−0.397*** (0.110)	−0.365** (0.107)	−0.424*** (0.109)
Percent Employed Construction	−0.851*** (0.203)	−0.855*** (0.200)	−0.813*** (0.182)
Percent Employed Wholesale	−0.647 (0.364)	−0.715 (0.371)	−0.723* (0.352)
Percent Employed Retail	−0.340** (0.115)	−0.314** (0.112)	−0.430*** (0.119)
Percent Employed Transportation	−0.232 (0.176)	−0.230 (0.173)	−0.259 (0.166)
Percent Employed IT	−0.260 (0.371)	−0.243 (0.360)	−0.290 (0.387)
Percent Employed Finance	−0.406** (0.125)	−0.410** (0.125)	−0.385*** (0.109)
Percent Employed Educational	−0.221* (0.099)	−0.219* (0.097)	−0.242* (0.104)
Percent Employed Professional	−0.200 (0.127)	−0.216 (0.127)	−0.186 (0.122)
Percent Employed Other Services	−0.399 (0.245)	−0.347 (0.245)	−0.373 (0.238)

Table 6.4 *Continued*

	Model 1	Model 2	Model 3
Percent Employed Public/	−0.435***	−0.425***	−0.428***
Government	(0.122)	(0.118)	(0.108)
Percent High School Degree	−0.038	−0.026	0.031
	(0.118)	(0.118)	(0.122)
Percent College Degree	−0.208**	−0.207**	−0.199**
	(0.063)	(0.062)	(0.063)
Percent Millennial	0.201*	0.200*	0.184*
	(0.081)	(0.080)	(0.080)
Percent Generation X	0.471***	0.471***	0.397**
	(0.124)	(0.120)	(0.116)
Percent Baby Boomers	0.417***	0.417***	0.352**
	(0.094)	(0.093)	(0.103)
Top 50% Counties for COVID-19		1.317*	
Cases		(0.510)	
Percent Low-Income Retail Jobs			0.281
Lost			(0.147)
Constant	10.503	9.078	8.444
	(9.970)	(10.047)	(10.139)
R-squared	0.369	0.378	0.407
N	2628	2628	2628

$*p < .05, **p < .01, ***p < .001$

subscription rate of 77.7%, would drop unemployment from 21.9% to 21.5%. The results are the same for metropolitan areas and counties, though the metro results are not shown here. Adding 1 site per 100 people to the metropolitan area of Detroit, Michigan, with a broadband subscription rate of 79.9%, would decrease unemployment from 24.4% to 24.0%

This analysis demonstrates the need to explore specific uses of broadband for their social and policy impacts. Examining broadband subscriptions alone would lead to the conclusion that internet use did not contribute to community resilience during the pandemic. The density of domain name sites alone was not enough to benefit communities in this particular case either, but having both a local context where many residents are online and the density of commercial websites did predict lower unemployment in communities. Company surveys have shown that only 8% of the domain name sites are for businesses with 11 or more employees, so this suggests that going online helped many small businesses during COVID-19. This was no small feat in the face of a sudden and severe economic crisis, which caused a "massive dislocation among small businesses" in the early months examined here (Bartik et al. 2020a, 2).

Future research on the impacts of broadband use for innovation and outcomes across geographies will inevitably need to utilize data generated by digital activity, including private data, as well as publicly available data on broadband subscriptions. The US Census Bureau's ACS with tract-level data is a step forward, but more is needed, including data on activities online. Along with the growing policy agenda for broadband, there is an unfinished research agenda as well.

Broadband Use and Policy for Opportunity

The diversity in federal systems offers the potential for creativity and innovation, as well as responsiveness to local needs. Federal systems can also foster inequality, and the policy choices we make will determine how to reconcile the two. Social inequality and economic outcomes are not solely due to markets, but shaped by these policy choices, for better or for worse (Bartels 2017; Franko and Witko 2017; Trounstine 2018).

Policies that are responsive to local needs and contexts require leadership by city and county governments, state and federal governments, and community-based organizations. Solutions for broadband adoption will differ in San Jose compared with Detroit, and in Wheeler County, Georgia, compared with Apache County, Arizona. Scholars have argued that there is a "new localism" (Katz and Nowak 2017) focused on the politics of problem solving and innovation (Katz and Nowak 2017; Barber 2013, xi), at a time when social change is taking place unevenly across communities, and when partisan gridlock has hampered policy solutions at other levels. But support is needed across levels of government and in the nonprofit and private sectors to develop effective solutions. This is especially the case for cities and counties with substantial low-income populations and high need. Whether urban or rural, communities with the most need may have the fewest local resources to change the calculus for the future.

Motivation to act on broadband and digital human capital has been strong in many local communities, and we provide evidence that there is good reason to promote this route to prosperity and growth. In contrast to winner-take-all competitions over corporate headquarters for Amazon or Google, digital human capital is not a zero-sum game. An earlier era of state and local economic development as smokestack chasing has shown that governments may shower unnecessarily generous benefits on companies they compete to attract, with little to show for the investment if firms depart (Jones and Bachelor 1993).

Digital human capital is a resource for a more inclusive approach to innovation. It is a resource that can promote growth across industries and occupations within communities, not just in the most skilled positions. Additionally, broadband use has the potential to improve public health, offer new educational resources, and more. It can be cultivated in a diversity of communities, creating smart cities or connecting remote places to the wider world. From the neighborhood to the state capital, broadband adoption represents access to information in a digital society, necessary for individuals and communities to realize their potential. The high human and economic costs of present inequalities are being recognized and will require intergovernmental and cross-sectoral solutions. Our analyses have shown that growth and prosperity increase as broadband use becomes more widespread, promoting skills and information throughout a community. It is an inclusive path toward innovation, and a 21st-century platform for opportunity.

Appendix A
Broadband Questions, Current Population Survey (CPS) 2000–2012/American Community Survey (ACS) 2013–2017

Broadband	2000 CPS Coded to 1 if *hescu8* = 2 and **access** = 1; 0, otherwise.	*hescu8*: Type of home internet access 1 Regular, or "dial-up," telephone service 2 Higher-speed internet access service
	2001 CPS, 2003 CPS Coded to 1 if *hesint2a* ≠ 1 and **access** = 1; 0, otherwise.	*hesint2a*: Do you currently access the internet using – 1 A regular "dial-up" telephone line 2 A DSL line (Digital Subscriber Line) 3 A cable modem 4 Something else

Appendix A Continued

Access to high-speed internet at home	2007 CPS, 2009 CPS Coded to 1 if *henet4* ≠ 1 and **access** = 1; 0, otherwise.	*henet4*: Do you currently access the Internet using – 1 A regular "dial-up" telephone 2 DSL, cable modem, satellite, wireless (such Wi-Fi), mobile phone or PDA, fiber-optics, or some other broadband Internet connection 3 Something else
	2010 CPS Coded to 1 if at least one variable from *heserv32*–*heserv37* is equal to 1 and **access** = 1; 0, otherwise.	*heserv32*: At home, (do you/does anyone in this household) access the internet using DSL service? *heserv33*: At home, (do you/does anyone in this household) access the internet using cable modem service? *heserv34*: At home, (do you/does anyone in this household) access the internet using fiber-optic service? *heserv35*: At home, (do you/does anyone in this household) access the internet using a mobile broadband plan for a computer or a cell phone? *heserv36*: At home, (do you/does anyone in this household) access the internet using satellite service? *heserv37*: At home, (do you/does anyone in this household) access the internet using some other service? 1 Yes 2 No

Appendix A Continued

| 1 = true | 2011 CPS
Coded to 1 if at least one variable from *hesci32–hesci37* is equal to 1 and **access** = 1;
0, otherwise. | *hesci72*: At home, does anyone in this household access the internet using DSL service?
hesci73: At home, does anyone in this household access the internet using cable modem service?
hesci74: At home, does anyone in this household access the internet using fiber-optic service?
hesci75: At home, does anyone in this household access the internet using a mobile broadband plan (for a computer or a cell phone)?
hesci76: At home, does anyone in this household access the internet using satellite service?
hesci77: At home, does anyone in this household access the internet using some other service?
1 Yes
2 No |
| | 2012 CPS
Coded to 1 if at least one variable from *henet42–henet47* is equal to 1 and **access** = 1;
0, otherwise. | *henet42*: At home, does anyone in this household access the internet using a DSL service?
henet43: At home, does anyone in this household access the internet using a cable modem service?
henet44: At home, does anyone in this household access the internet using a fiber-optic service?
henet45: At home, does anyone in this household access the internet using a mobile broadband plan (for a computer, cell phone, smart phone, or tablet)?
henet46: At home, does anyone in this household access the internet using a satellite service?
henet47: At home, does anyone in this household access the internet using some other service?
1 Yes
2 No |

Continued

Appendix A Continued

0 = other	2013–2017 ACS	*cimodem*: cable internet service
	Coded to 1 if at least	*cisat*: satellite internet service
	one variable from	*cids1*: DSL service
	cimodem, cisat, cidsl,	*cifiber*: fiber-optic internet service
	cifiber, cibrdbnd, and	*cibrdbnd*: mobile broadband plan
	ciothsvc is equal to 1;	*ciothsvc*: other internet service
	0, otherwise.	0 N/A
		1 Yes
		2 No

Notes

Chapter 1

1. The Organisation for Economic Cooperation and Development (OECD) publishes Akamai's average speed for member countries. Data for Q1 2019 shows the United States with 18.7 Mbps on average. One gigabit is 125 megabits (Mb).
2. https://www.census.gov/quickfacts/fact/table/rosebudcountymontana
3. The source of data for many previous studies is Form 477 from the FCC, data that is problematic in many ways. The broadband deployment data has been criticized for overstating availability (counting a census tract as served if even just one household can be connected) (Grubesic 2012). Areas with between one and three service providers are also reported as the same, for reasons of confidentiality. Despite these limitations, the FCC deployment data is often used in research because of its availability for census tracts and over time.
4. University of Iowa-Arizona State University Broadband Data Portal supported by National Science Foundation grant #1338471. https://techdatasociety.asu.edu/broadband-data-portal/home
5. Longitudinal data presented here covers only the 820 largest counties (out of 3,100 in the United States).
6. Other sociologists and communications scholars have used the concept of digital capital, which Ragnedda (2018) defines as access, skills, and outcomes (see also Park 2017; Scheerder et al. 2017 for discussion of digital capital). This is usually defined on an individual level, but Bach and colleagues (2013) make the connection to human capital in communities, like educational attainment. We extend this observation by more explicitly examining the links between this digital human capital and community gains.
7. The report called the informational and strategic skills "navigational," though the latter term has been used differently by other authors (van Deursen and van Dijk 2011).
8. See Tolbert and Mossberger (2015, Part 7) on technical documentation for question wording in the CPS surveys.

Chapter 2

1. The US Census Bureau publishes County Business Patterns, for example, and this has also been the level at which many studies of broadband infrastructure and availability have been conducted.
2. https://www.douglas.co.us/about-us/douglas-county-history/

3. https://www.co.apache.az.us/home-page/
4. Data on adoption was available from the Federal Communications Commission prior to 2018 but was a blunt instrument for measuring differences over places or time, given that all data was aggregated in quintiles.
5. See also Whitacre et al. (2014b) for similar results using different methods of analysis.
6. We define age cohorts by the following years: Silent (1945 or earlier), Baby Boomer (1946–1965), Generation X (1965–1980), Millennial (1981–1996), and Post-Millennial/Gen Z (after 1996).
7. We have estimated replications for median income in Tables 2.2. through 2.4 using median income logged, and the substantive findings are unchanged.

Chapter 3

1. A binary year variable for 2015 is included given a small systematic decline in broadband subscriptions for this year only, likely due to a sampling issue with the census data as the question wording did not change. Exclusion of the year binary variable does not change the results reported here, but because it is an outlier in the time series, it is controlled for.

Chapter 4

1. Even consolidated city-county governments, like Miami-Dade County or Louisville Metro, seldom cover the entire metropolitan region. And these consolidated governments are very much the exception rather than the rule in the United States. Only 40 out of the more than 3,000 counties in the United States are consolidated (https://www.nlc.org/resource/cities-101-consolidations).
2. ACS (2017)—broadband of all types was estimated at 78.1% for the United States, and cell phone–only internet use was 7.5%, included in this total. The 2017 five-year averages are used here for consistency with the zip code–level data, which is not available in the one-year estimates. The five-year averages are a little lower than the 2017 one-year estimates in the ACS. Broadband subscriptions of all types for the United States are 83.8% in the 2017 one-year estimates, compared to 78.1% for 2017.
3. https://docs.fcc.gov/public/attachments/DOC-353962A1.pdf
4. For the first time, five-year estimates for the ACS became available at the census tract level in December 2018. For smaller samples, such as census tracts (or for demographic groups within cities), the ACS provides only five-year estimates. The ACS estimates are a much better measure of differences across places than 20-percentage-point categories in the FCC data.
5. Zip code data can be problematic in rural areas or for comparisons over time (Grubesic and Matisziw 2006), but these are not relevant for their use in this chapter.

6. http://www.seattle.gov/tech/initiatives/digital-equity/technology-matching-fund
7. http://www.seattle.gov/tech/initiatives/digital-equity/digital-equity-progress-report
8. http://www.seattle.gov/tech/initiatives/digital-equity/technology-matching-fund/2018-awardees
9. http://austintexas.gov/sites/default/files/files/Digital_Inclusion_Strategy_ADOPTED.pdf
10. http://austintexas.gov/sites/default/files/files/Telecommunications/DigitalInclusion/Digital_Inclusion_Business_Plan_With_Roadmap_Report.pdf
11. http://austintexas.gov/page/digital-inclusion-strategic-plan
12. http://austintexas.gov/page/digital-empowerment-community-austin
13. http://austintexas.gov/digitalinclusion
14. http://austintexas.gov/page/community-connections-program
15. http://austintexas.gov/department/grant-technology-opportunities-program
16. https://www.portlandoregon.gov/oct/article/643857; Digital Equity Action Plan, 4.
17. https://www.portlandoregon.gov/oct/73863
18. https://www.portlandoregon.gov/oct/76093
19. https://www.portlandoregon.gov/oct/73859
20. https://www.boston.gov/departments/broadband-and-cable/broadband-and-digital-equity#our-work
21. http://www.mayorsfundphila.org/initiatives/digital-literacy-alliance/
22. https://detroitmi.gov/node/20466
23. 313 is Detroit's area code.
24. https://www.nola.com/politics/2018/08/tell_cantrell_digital_equity.html
25. The surveys were conducted by the Eagleton Institute at Rutgers University. The random sample telephone surveys used a geographic sampling frame that drew respondents from each of Chicago's 77 community areas in a stratified sample. The interviews were conducted in English and Spanish.
26. This method creates geographic estimates of critical outcome variables but leverages the neighborhood-level socioeconomic data to improve estimates based on individual-level data. This method has been shown to work well with a small number of cases in each geographic area (Lax and Phillips 2009a, 2009b; Raudenbush and Bryk 2002; Snijders and Bosker 2012; Steenbergen and Jones 2002).
27. The data were downloaded from the US Census Bureau website at the tract level and then aggregated to the neighborhood level, weighted by community area population size. As with the dependent variables, the independent variables used in this analysis are the differences between the 2008 and 2013 neighborhood-level values.
28. Statistically significant with 95% confidence interval for a direction (one-tailed) significance test.
29. In column 2 of Table 1A (Mossberger, Tolbert and Anderson 2014) the coefficient for the treatment (Chicago Smart Communities) is again positive and statistically significant for home broadband adoption, with the 92% confidence interval (or 96% confidence interval for a directional test).
30. The coefficient for the treatment (Chicago Smart Communities) is positive and statistically significant with a 95% confidence interval.

Chapter 5

1. As Justice Louis Brandeis referred to the states in a 1932 Supreme Court decision, *New State Ice Co. v. Liebmann.*
2. This chapter is partially based on an earlier article that appeared in Political Research Quarterly, LaCombe, S., Tolbert, C.J., & Mossberger, K. (2021), Information and Policy Innovation in US States, https://doi.org/10.1177/10659129211006783
3. While single-policy studies have often found that legislative professionalism matters (Berry 1994; Shipan and Volden 2006), prior analyses of multiple policies (Boehmke and Skinner 2012) have found that legislative professionalism is not associated with policy adoptions.

Chapter 6

1. Navajo Nation reservation and off-reservation trust land, New Mexico–Arizona–Utah. Sprawling across three states, the Navajo Nation is larger than nine of the US states (National Congress of American Indians 2020) and has over 166,000 residents who are classified by the census as American Indian/Alaska Native (AI/AN). ACS 2017 five-year estimates, Tables B28009 and B28009C.
2. https://www.census.gov/newsroom/press-releases/2018/2013-2017-acs-5year.html
3. Three focus groups of smartphone-dependent internet users were conducted by researchers at the University of Illinois at Chicago (UIC) in 2011–2012, with support from the UIC Institute for Policy and Civic Engagement. There were six participants in the English-language group and 10 in the two Spanish-language groups. Overall, respondents were mixed in terms of race, ethnicity, gender, and age (from college students to retirees). See also Brown et al. (2012) for a more detailed report on the first two groups.
4. https://www.fcc.gov/document/fcc-launches-20-billion-rural-digital-opportunity-fund
5. https://www.usda.gov/broadband
6. The exceptions are discounted programs through Google Fiber (25 Mbps) and Spectrum (30 Mbps).
7. See also Mossberger et al. (2003, Chapter 3) on the skills divide.
8. The March 2020 legislation provided $150 billion to state, local and Tribal governments for Covid-related expenses, and many state and local governments used the funds for devices and connections for students, public Wi-Fi, telehealth, and improved broadband infrastructure (deWit 2020). The $81.8 billion for the Governors' Emergency Education Relief Fund in the December 2020 bill allowed spending to support school and student internet use as well (Benton Institute 2021a). Another $10 billion allocated for capital funds projects in March 2021 and the $219.8 billion for state, local and Tribal governments can be used for broadband infrastructure, among other purposes (Benton Institute 2021c).

9. https://www.pewtrusts.org/en/research-and-analysis/data-visualizations/2019/state-broadband-policy-explorer
10. Task forces study issues and make recommendations, while councils, which generally are more involved in overseeing programs, may be used to span state agencies or to encourage public-private partnerships (Stauffer et al. 2020).
11. https://www.digitalinclusion.org/blog/2019/06/18/ndia-cwa-public-knowledge-file-brief-in-case-vs-fccs-5g-preemption/
12. https://www.grow.google/
13. https://www.godaddy.com/godaddy-for-good
14. https://news.microsoft.com/rural-broadband/
15. https://www.nytimes.com/2020/04/05/opinion/coronavirus-google-searches.html
https://theconversation.com/sudden-loss-of-smell-why-it-is-a-reason-to-self-isolate-138682
16. http://i.stanford.edu/~julian/pdfs/nips2012.pdf

References

Abraham, T. (2015). What Other Cities Should Learn from Philly's Failed Municipal Broadband Effort. *Technical.Ly*. https://technical.ly/philly/2015/03/04/cities-learn-phillys-failed-municipal-broadband-effort/

Abrardi, L., & Cambini, C. (2019). Ultra-Fast Broadband Investment and Adoption: A Survey. *Telecommunications Policy, 43*(3), 183–198.

Ahmad, Z. (2018). *Flint Turns to Tech to Try and Solve Crime, Opioid Problems*. https://www.mlive.com/news/flint/2018/04/smart_city_competition_in_flin.html

Akerman, A., Gaarder, I., & Mogstad, M. (2015). The skill complementarity of broadband internet. *Quarterly Journal of Economics, 130*(4), 1781–1824.

Albino, V., Berardi, U., & Dangelico, R. M. (2015). Smart Cities: Definitions, Dimensions, and Performance. *Journal of Urban Technology, 22*(1), 3–21.

Allard, S. (2017). *Places in Need: The Changing Geography of Poverty*. Russell Sage.

Anderson, M. (2019). *Mobile Technology and Home Broadband 2019*. https://www.pewinternet.org/2019/06/13/mobile-technology-and-home-broadband-2019/

Angrist, J. D., & Pischke, J. S. (2009). *Mostly Harmless Econometrics: An Empiricist's Companion*. Princeton University Press.

Applebome, P. (2016). In Detroit's 2-Speed Recovery, Downtown Roars and Neighborhoods Sputter. *New York Times*. https://www.nytimes.com/2016/08/13/us/detroit-recovery.html

Associated Press. (2020). School Shutdowns Raise Stakes of Digital Divide for Students. *New York Times*. https://www.nytimes.com/aponline/2020/03/30/us/ap-us-virus-outbreak-digital-divide.html

Atasoy, H. (2013). The Effects of Broadband Internet Expansion on Labor Market Outcomes. *Industrial & Labor Relations Review, 66*(2), 315–345.

Atkinson, K., & Nilles, K. (2008). *Tribal Business Structure Handbook*. Office of Indian Energy and Economic Affairs. https://www.irs.gov/pub/irs-tege/tribal_business_structure_handbook.pdf

Auxier, B., & Anderson, M. (2020). *As Schools Close Due to the Coronavirus, Some U.S. Students Face a Digital "Homework Gap."* Pew Research Center. https://www.pewresearch.org/fact-tank/2020/03/16/as-schools-close-due-to-the-coronavirus-some-u-s-students-face-a-digital-homework-gap/

Bach, A., Shaffer, G., & Wolfson, T. (2013). Digital Human Capital: Developing a Framework for Understanding the Economic Impact of Digital Exclusion in Low-Income Communities. *Journal of Information Policy, 3*, 247–266.

Badger, E. (2018). In Superstar Cities, the Rich Get Richer and They Get Amazon. *New York Times*. https://www.nytimes.com/2018/11/07/upshot/in-superstar-cities-the-rich-get-richer-and-they-get-amazon.html

Balboa, E. (2020). *Rocket Mortgage Classic Announces Fund to Tackle Detroit's Tech Access, Digital Literacy Gaps*. Benzinga. https://www.benzinga.com/news/20/06/16349912/

rocket-mortgage-classic-announces-fund-to-tackle-detroits-tech-access-digital-literacy-gaps

Balla, S. J. (2001). Interstate Professional Associations and the Diffusion of Policy Innovations. *American Politics Research, 29*(3), 221–245.

Ban, C. (2018). *Broadband Serves as a Rural Lifeline and a Building Block.* National Association of Counties (NACo). https://www.naco.org/articles/broadband-serves-rural-lifeline-and-building-block.

Barber, B. R. (2013). *If Mayors Ruled the World: Dysfunctional Nations, Rising Cities.* Yale University Press.

Barker, R., & Lanham, D. (2018). *The Fast Lane—Amazon's Announcement, Understanding the Heartland, and White Voter Shift to Democrats.* Brookings Institution. https://www.brookings.edu/blog/the-avenue/2018/11/16/the-fast-lane-amazons-announcement-understanding-the-heartland-and-white-voter-shift-to-democrats/

Bartels, L. M. (2016). *Unequal Democracy: The Political Economy of the New Gilded Age.* 2nd ed. Princeton University Press.

Bartels, L. M. (2017). *Political Inequality in Affluent Democracies: The Social Welfare Deficit* (No. 5–2017). Center for the Study of Democratic Institutions, Vanderbilt University. https://www.vanderbilt.edu/csdi/research/Working_Paper_5_2017.pdf

Bartik, A. W., Bertrand, M., Cullen, Z. B., Glaeser, E. L., Luca, M., & Stanton, C. T. (2020a). *How are small businesses adjusting to covid-19? Early evidence from a survey* (No. w26989). National Bureau of Economic Research. https://www.nber.org/system/files/working_papers/w26989/w26989.pdf

Bartik, A. W., Bertrand, M., Liu, F., Rothstein, J., & Unrath, M. (2020b). Measuring the Labor Market at the Onset of the COVID-19 Crisis. Brookings Papers on Economic Activity, June 25, 2020. https://www.brookings.edu/wp-content/uploads/2020/06/Bartik-et-al-conference-draft.pdf

Baumgartner, F., & Jones, B. D. (2005). *Agendas and Instability in American Politics.* University of Chicago Press.

Beck, N., & Katz, J. N. (1996). Nuisance Vs. Substance: Specifying and Estimating Time-Series-Cross-Section Models. *Political Analysis: An Annual Publication of the Methodology Section of the American Political Science Association, 6*(1), 1–36.

Becker, G. S. (1964). *Human Capital: A Theoretical and Empirical Analysis, with Special Reference to Education.* Columbia University Press.

Bekiempis, V. (2020). Pandemic Response Lays Bare America's Digital Divide. *The Guardian.* https://www.theguardian.com/world/2020/mar/21/coronavirus-us-digital-divide-online-resources

Benda, N. C., Veinot, T. C., Sieck, C. J., & Ancker, J. S. (2020). Broadband Internet Access Is a Social Determinant of Health! *American Journal of Public Health, 110*(8), 1123–1125.

Benton Institute. (2020). *How Does the CARES Act Connect Us?* https://www.benton.org/blog/how-does-cares-act-connect-us

Benton Institute. (2021a). *The Last Broadband Gifts from the 116th Congress.* https://www.benton.org/blog/last-broadband-gifts-116th-congress

Benton Institute. (2021b). *Introducing the Emergency Broadband Benefit Program.* https://www.benton.org/blog/introducing-emergency-broadband-benefit-program

Benton Institute. (2021c). *American Rescue Plan: Broadband and the Social Safety Net.* https://www.benton.org/blog/american-rescue-plan-broadband-and-social-safety-net

Bergman, P., Chetty, R., DeLuca, S., Hendren, N., Katz, L. F., & Palmer, C. (2019). *Creating Moves to Opportunity: Experimental Evidence on Barriers to Neighborhood Choice* (No. 26164). NBER Working Paper. https://opportunityinsights.org/paper/cmto

Bergson-Shilcock, A. (2020). *The New Landscape of Digital Literacy*. National Skills Coalition. https://www.nationalskillscoalition.org/resources/publications/file/New-Digital-Landscape-web.pdf?emci=3d23029b-5baa-ea11-9b05-00155d039e74&emdi=ea000000-0000-0000-0000-000000000001&ceid=5726379

Berry, F. S. (1994). Sizing Up State Policy Innovation Research. *Policy Studies Journal, 22*(3), 442–456.

Berry, W. D., & Baybeck, B. (2005). Using Geographic Information Systems to Study Interstate Competition. *American Political Science Review, 99*(4), 505–519.

Berry, F. S., & Berry, W. D. (1990). State Lottery Adoptions as Policy Innovations: An Event History Analysis. *American Political Science Review, 84*(2), 395–415.

Bertschek, I., Briglauer, W., Hüschelrath, K., Kauf, B., & Niebel, T. (2016). *The Economic Impacts of Telecommunications Networks and Broadband Internet: A Survey* (No. 16–056). http://ftp.zew.de/pub/zew-docs/dp/dp16056.pdf

Berube, A. (2018). *For Amazon, HQ2 Location Decision Was about Talent, Talent, Talent*. Brookings Institution. https://www.brookings.edu/blog/the-avenue/2018/11/13/for-amazon-hq2-location-decision-was-about-talent-talent-talent/

Berube, A., & Murray, C. (2018). *Renewing America's Economic Promise through Older Industrial Cities*. Metropolitan Policy Program at Brookings.

Bischoff, K., & Reardon, S. F. (2014). Residential Segregation by Income, 1970–2009. In J. R. Logan (Ed.), *Diversity and Disparities: America Enters a New Century* (pp. 208–233). Russell Sage Foundation.

Boehmke, F. J., Brockway, M., Desmarais, B. A., Harden, J. J., LaCombe, S., Fridolin, L., & Wallach, H. (2020). SPID: A New Database for Inferring Public Policy Innovativeness and Diffusion Networks. *Policy Studies Journal, 48*(2), 517–545. https://doi.org/https://doi.org/10.1111/psj.12357

Boehmke, F. J., Desmarais, B. A., Harden, J. J., Wallach, H., LaCombe, S., Fridolin, L., & Wallace, D. (2019). SPID: State Policy Innovation and Diffusion Database. https://dataverse.harvard.edu/dataverse/spid

Boehmke, F. J., & Skinner, P. (2012). State Policy Innovativeness Revisited. *State Politics & Policy Quarterly, 12*(3), 303–329.

Boorsma, B. (2018). *A New Digital Deal: Beyond Smart Cities. How to Best Leverage Digitalization for the Benefit of Our Communities*. Rainmaking Publications.

Bouchet, M., Liu, S., Parilla, J., & Kabbani, N. (2018). *Global Metro Monitor 2018*. https://www.brookings.edu/Research/Global-Metro-Monitor-2018/

Boudette, N.E. (2018). Ford Aims to Revive a Detroit Train Station, and Itself. *New York Times*, June 17, 2018. https://www.nytimes.com/2018/06/17/business/ford-detroit-station.html

Boushey, G. (2010). *Policy Diffusion Dynamics in America*. Cambridge University Press.

Box-Steffensmeier, J. M., & Jones, B. S. (2004). *Event History Modeling: A Guide for Social Scientists (Analytical Methods for Social Research)*. Cambridge University Press.

Broadband Data Portal. Iowa-ASU Broadband Data Portal Home. https://techdatasociety.asu.edu/broadband-data-portal/home

Broadband Now. (2020). Municipal Broadband Is Roadblocked or Outlawed in 22 States. https://broadbandnow.com/report/municipal-broadband-roadblocks/

Brookings Institution & Federal Reserve. (2008). *The Enduring Challenge of Concentrated Poverty in America: Case Studies from across the U.S.* https://www.brookings.edu/wp-content/uploads/2016/06/1024_concentrated_poverty.pdf

Brown, A., Benoit Bryan, J., & Mossberger, K. (2012). *Smartphones and Minorities: Closing Gaps or Creating New Disparities?* Presented at the Annual Meeting of the Midwest Political Science Association, April 2012, Chicago.

Brown, M. (2019). Tesla World Map Shows EV Hot Spots in Cities across the Globe. *Inverse.* https://www.inverse.com/article/55170-tesla-s-fleet-map

Brunston, S. (2020). *FCC: Don't Abandon Tribes during a Pandemic, Extend the 2.5 GHz Rural Tribal Priority Window.* https://www.publicknowledge.org/blog/fcc-dont-abandon-tribes-during-a-pandemic-extend-the-2-5-ghz-rural-tribal-priority-window/

Brynjolfsson, E., & Saunders, A. (2010). *Wired for Innovation: How Information Technology Is Reshaping the Economy.* MIT Press.

Butler, D. M., Volden, C., Dynes, A. M., & Shor, B. (2017). Ideology, Learning, and Policy Diffusion: Experimental Evidence. *American Journal of Political Science, 61*(1), 37–49. https://doi.org/10.1111/ajps.12213

California Emerging Technology Fund (CETF). (n.d.). Public Awareness and Education. https://www.cetfund.org/action-and-results/public-awareness-and-education-get-connected/.

California Public Utilities Commission. (2020). *California Advanced Services Fund Adoption Account.* https://www.cpuc.ca.gov/General.aspx?id=6442457502.

Callahan, B. (2019a). AT&T's Digital Redlining of Dallas: New Research by Dr. Brian Whitacre. Benton Institute for Broadband and Society. https://www.benton.org/headlines/att's-digital-redlining-dallas-new-research-dr-brian-whitacre

Callahan, B. (2019b). NDIA, CWA, Public Knowledge File Brief in Case vs. FCC's 5G Preemption. https://www.digitalinclusion.org/blog/2019/06/18/ndia-cwa-public-knowledge-file-brief-in-case-vs-fccs-5g-preemption/

Callahan, B. (2019c). Representatives McNerney, Lujan and Clarke Introduce Digital Equity Act in U.S. House. https://www.digitalinclusion.org/blog/2019/09/25/representatives-mcnerney-lujan-and-clarke-introduce-digital-equity-act-in-u-s-house/

Callahan, B. (2020). *U.S. House Votes "Yes" on Home Broadband Benefit and Digital Equity Funding for States and Communities.* National Digital Inclusion Alliance. https://www.digitalinclusion.org/blog/author/bill/

Caplovitz, D. (1967). *The Poor Pay More: Consumer Practices of Low-Income Families.* Free Press.

Caughey, D., & Warshaw, C. (2016). The Dynamics of State Policy Liberalism, 1936–2014. *American Journal of Political Science, 60*(4), 899–913. https://doi.org/https://doi.org/10.1111/ajps.12219

Center on Rural Innovation. (n.d.). *The RII Community Toolkit.* https://ruralinnovation.us/rural-innovation-initiative/public-toolkit/

Chadwick, A. (2017). *The Hybrid Media System: Politics and Power.* Oxford University Press.

Chao, B., & Park, C. (2020). *The Cost of Connectivity.* Open Technology Institute, New America Foundation. https://d1y8sb8igg2f8e.cloudfront.net/documents/The_Cost_of_Connectivity_2020__XatkXnf.pdf

Chetty, R., Friedman, J. N., Hendren, N., Jones, M. R., & Porter, S. R. (2018). *The Opportunity Atlas: Mapping the Childhood Roots of Social Mobility* (No. 25147). National Bureau of Economic Research.

Chetty, R., Friedman, J. N., Hendren, N., Stepner, M., & The Opportunity Insights Team. (2020). *The Economic Impacts of COVID-19: Evidence from a New Public Database Built Using Private Sector Data.* https://opportunityinsights.org/wp-content/uploads/2020/05/tracker_paper.pdf

Chetty, R., & Hendren, N. (2017). The Impacts of Neighborhoods on Intergenerational Mobility II: County-Level Estimates. *Quarterly Journal of Economics, 133*(3), 1163–1228. http://www.equality-of-opportunity.org/assets/documents/movers_paper2.pdf

Cobb, R. W., & Elder, C. D. (1972). *Participation in American Politics: The Dynamics of Agenda-Building.* Allyn and Bacon.

Comcast. (n.d.). Internet Essentials. https://corporate.comcast.com/values/internet-essentials

Community Networks. (2019). The State of State Preemption, Nineteen Is the Number—Community Broadband Bits Podcast 368. https://muninetworks.org/content/the-state-of-state-preemption-nineteen-is-the-number-community-broadband-bits-podcast-368

Community Networks. (2020). Community Network Map. https://muninetworks.org/communitymap

Community Networks. (n.d.). Fact Sheets. Next Century Cities. The Opportunity of Municipal Broadband. https://ilsr.org/wp-content/uploads/2019/09/2019-09-fact-sheet-NCC-opportunity-municipal-broadband.pdf

Crandall, R., Lehr, W., & Litan, R. (2007). *The Effects of Broadband Deployment on Output and Employment: A Cross-Sectional Analysis of US Data.* Brookings Institution. https://www.brookings.edu/research/the-effects-of-broadband-deployment-on-output-and-employment-a-cross-sectional-analysis-of-u-s-data/

Crawford, S. (2018). *Fiber: The Coming Tech Revolution—And Why America Might Miss It.* Yale University Press.

Cuyahoga County. (2019). County Partners with Digital C on Pilot to Provide Fairfax Neighborhood Low-Cost Internet Access. http://executive.cuyahogacounty.us/en-US/DigitalC-Home-Internet-Access.aspx

Dailey, D., Bryne, A., Powell, A., Karaganis, J., & Chung, J. (2010). *Broadband Adoption in Low-Income Communities.* https://pdfs.semanticscholar.org/dde5/13083ed19e7187e833cd41ae421a35d7064c.pdf

Descant, S. (2019). Philadelphia Adopts Smart City Road Map. *Government Technology.* https://www.govtech.com/fs/infrastructure/Philadelphia-Adopts-Smart-City-Road-Map.html

Desmarais, B. A., Harden, J. J., & Boehmke, F. J. (2015). Persistent Policy Pathways: Inferring Diffusion Networks in the American States. *American Political Science Review, 109*(2), 392–406.

Detroit Regional Chamber. (n.d.). Industry Clusters. https://www.detroitchamber.com/econdata/data/industry-clusters/

Dettling, L. J., Goodman, S., & Smith, J. (2018). Every Little Bit Counts: The Impact of High-Speed Internet on the Transition to College. *Review of Economics and Statistics, 100*(2), 260–273.

deWit, K. 2020. States Tap Federal CARES Act to Expand Broadband. November 16, 2020. Pew Charitable Trusts. https://www.pewtrusts.org/en/research-and-analysis/issue-briefs/2020/11/states-tap-federal-cares-act-to-expand-broadband

DiMaggio, P., & Bonikowski, B. (2008). Make Money Surfing the Web? The Impact of Internet Use on the Earnings of US Workers. *American Sociological Review, 73*(2), 227–250.

DiMaggio, P., Hargittai, E., Neuman, R. W., & Robinson, J. P. (2001). Social Implications of the Internet. *Annual Review of Sociology, 27*(1), 307–336.

Dobson, J. E., & Campbell, J. S. (2014). The Flatness of US States. *Geographical Review, 104*(1), 1–9.

Douglas, T. (2018, June). San Jose's Telecom Pacts Expand Broadband Infrastructure, Digital Equity. *Government Technology*. https://www.govtech.com/network/San-Joses-Telecom-Pacts-Expand-Broadband-Infrastructure-Digital-Equity.html

Duarte, M. E. (2017). *Network Sovereignty: Building the Internet across Indian Country.* University of Washington Press.

Durbin, D. (2019). Durbin, Maloney Introduce Bicameral Bill to Increase Access to Broadband Service for Low-Income Americans. https://www.durbin.senate.gov/news-room/press-releases/durbin-maloney-introduce-bicameral-bill-to-increase-access-to-broadband-service-for-low-income-americans

Economic Innovation Group. (2018). *From Great Recession to Great Reshuffling: Charting a Decade of Change across American Communities.* https://eig.org/dci

Efird, L. (2020). *Rural Broadband—Yancey and Mitchell Counties.* https://ncimpact.sog.unc.edu/2020/07/rural-broadband-yancey-and-mitchell-counties/

EveryoneOn. (n.d.). EveryoneOn. https://www.everyoneon.org/about-us#history

Eyestone, R. (1977). Confusion, Diffusion, and Innovation. *American Political Science Review, 71*(2), 441–447.

Falck, O. (2017). *Does Broadband Infrastructure Boost Employment?* IZA World of Labor.

Federal Communications Commission. (2010). *Connecting America: The National Broadband Plan.* https://transition.fcc.gov/national-broadband-plan/national-broadband-plan.pdf

Federal Communications Commission. (2018a). *E-Rate: Universal Service Program for Schools and Libraries.* https://www.fcc.gov/consumers/guides/universal-service-program-schools-and-libraries-e-rate

Federal Communications Commission. (2018b). *FCC Fact Sheet.* https://docs.fcc.gov/public/attachments/DOC-353962A1.pdf

Federal Communications Commission. (2018c). *FCC Takes Steps to Transform the 2.5 GHz Band for Next Generation 5G Connectivity.* https://docs.fcc.gov/public/attachments/DOC-350646A1.pdf

Federal Communications Commission. (2019). *2019 Broadband Deployment Report.* https://docs.fcc.gov/public/attachments/FCC-19-44A1.pdf

Federal Communications Commission. (2020a). *Fixed Broadband Deployment.* https://broadbandmap.fcc.gov/

Federal Communications Commission. (2020b). *Keep Americans Connected.* https://www.fcc.gov/keep-americans-connected

Feld, H. (2019). Solving the Rural Broadband Equation at the Local Level. *State and Local Government Review, 51*(4), 242–249.

Fernandez, L., Reisdorf, B. C., & Dutton, W. H. (2020). Urban Internet Myths and Realities: A Detroit Case Study. *Information, Communication and Society, 23*(13), 1925–1946.

Florida, R. (2017). *The New Urban Crisis: How Our Cities Are Increasing Inequality, Deepening Segregation, and Failing the Middle Class—And What We Can Do about It.* Basic Books.

Florida, R. (2019). America's Tech Hubs Still Dominate, but Some Smaller Cities Are Rising. *CityLab.* https://www.citylab.com/life/2019/04/tech-jobs-posting-by-city-hubs-indeed-employment-data-map/587347/

Florida, R., & Pedigo, S. (2020). *How Our Cities Can Reopen after the COVID-19 Pandemic.* Brookings Institution. https://www.brookings.edu/blog/the-avenue/2020/03/24/how-our-cities-can-reopen-after-the-covid-19-pandemic/

Forman, C., Goldfarb, A., & Greenstein, S. (2005). *Technology Adoption in and Out of Major Urban Areas: When Do Internal Firm Resources Matter Most?* (No. 11642). https://www.nber.org/papers/w11642.pdf

Forman, C., Goldfarb, A., & Greenstein, S. (2008). Understanding the Inputs into Innovation: Do Cities Substitute for Internal Firm Resources? *Journal of Economics & Management Strategy, 17*(2), 295–316.

Forman, C., Goldfarb, A., & Greenstein, S. (2012). The Internet and Local Wages: A Puzzle. *American Economic Review, 102*(1), 556–575.

Franko, W., & Witko, C. (2017). *The New Economic Populism: How States Respond to Economic Inequality.* Oxford University Press.

Frey, C. B., Ainley, J., Curmi, E., Garlick, R., Ierodiaconou, G., Pejaver, N., Pollard, M., Singlehurst, T. A., Chen, C., Bilerman, M., Galla, A., Guy, A., Manduca, M. A., Ping, J., & Pritchard, W. H. (2020). *Technology at Work: A New World of Remote Work.* Citi GPS: Global Perspectives & Solutions. https://ir.citi.com/td2TMf%2FvvpzNPqaucEszMhDfq%2Fq%2ByImXWvzH61WVNip7Ecd1v7edrIrz6nCHdxkoR2AmAYyMDa4%3D

Frey, W. H. (2018). *Where Do the Most Educated Millennials Live? Are They Living in the Amazon HQ2 Finalist Places?* https://www.brookings.edu/blog/the-avenue/2018/02/06/where-do-the-most-educated-millennials-live/

Gallardo, R., Whitacre, B., Kumar, I., & Upendram, S. (2021). Broadband Metrics and Job Productivity: A Look at County-Level Data. *Annals of Regional Science, 66*, 261–184.

Galperin, H., Bar, F., Kim, A. M., Bui, M., & Li, X. (2017). *Mapping Digital Exclusion in Los Angeles County.* http://arnicusc.org/wp-content/uploads/2017/07/Policy-Brief-2.pdf

Galster, G. (2017). Our Metro Areas Have Become Engines of Inequality. *Urban Affairs Review Forum.* https://urbanaffairsreview.com/2017/01/24/our-metro-areas-have-become-engines-of-inequality/

Gandara, D., Rippner, J. A., & Ness, E. C. (2017). Exploring the "How" in Policy Diffusion: National Intermediary Organizations' Roles in Facilitating the Spread of Performance-Based Funding Policies in the States. *Journal of Higher Education, 88*(5), 701–725.

Gangadharan, S. P., and Byrum, G. (2012). Introduction: Defining and Measuring Meaningful Broadband Adoption. *International Journal of Communication, 6*, 2601–2608.

Garrett, K. N., & Jansa, J. M. (2015). Interest Group Influence in Policy Diffusion Networks. *State Politics & Policy Quarterly, 15*(3), 387–417.

Gaskell, A. (2020). Are We Entering a New World of Remote Work? *Forbes*, July 8. https://.
www.forbes.com/sites/adigaskell/2020/07/08/are-we-entering-a-new-world-of-
remote-work/?sh=7b91d42e5959

Giannone, E. (2017). Skill-Biased Technical Change and Regional Convergence. In *2017
Meeting Papers* (No. 190). Society for Economic Dynamics.

Gilardi, F. (2016). Four Ways We Can Improve Policy Diffusion Research. *State Politics &
Policy Quarterly*, 16(1), 8–21.

Gillett, S., Lehr, W. H., Osorio, C. A., & Sirbu, M. A. (2006). *Measuring Broadband's
Economic Impact*. Economic Development Administration.

Gilroy, A. A. (2017). *Federal Lifeline Program: Frequently Asked Questions*. Congressional
Research Service. https://purl.fdlp.gov/GPO/gpo112421.

Glaeser, E. (2011). *Triumph of the City: How Our Greatest Invention Makes Us Richer,
Smarter, Greener, Healthier, and Happier*. Penguin Press.

Goldberg, R. (2019). *Unplugged: NTIA Survey Finds Some Americans Still Avoid Home
Internet Use*. National Telecommunications and Information Administration. https://
www.ntia.doc.gov/blog/2019/unplugged-ntia-survey-finds-some-americans-
still-avoid-home-internet-use

Goldin, C. (1998). America's Graduation from High School: The Evolution and Spread
of Secondary Schooling in the Twentieth Century. *Journal of Economic History*, 58(2),
345–374.

Goldsmith, S., & Crawford, S. (2014). *The Responsive City: Engaging Communities through
Data-Smart Governance*. Jossey-Bass.

Goldstein, P. (2020). How Cities Are Forging Partnerships to Close the Digital
Divide. StateTech, December 10. https://statetechmagazine.com/article/2020/12/
how-cities-are-forging-partnerships-close-digital-divide

Goodman, J. D. (2019). Amazon Pulls Out of Planned NYC Headquarters. *New York
Times*, February 14. https://www.nytimes.com/2019/02/14/nyregion/amazon-hq2-
queens.html

Goolsbee, A., & Klenow, P. J. (2002). Evidence on Learning and Network Externalities
in the Diffusion of Home Computers. *Journal of Law and Economics*, 45(2), 317–343.
https://doi.org/10.1086/344399

Government Accountability Office. (2018). *Broadband Internet: FCC's Data Overstate
Access on Tribal Lands*.

Government Technology. (2019, April 10). *Detroit Hires Its First Director of Digital
Inclusion*. https://www.govtech.com/network/Detroit-Hires-its-First-Director-of-
Digital-Inclusion.html

Grand-Clement, S. (2017). *Digital Learning: Education and Skills in the Digital Age*.
RAND Europe.

Gray, V. (1973). Innovation in the States: A Diffusion Study. *American Political Science
Review*, 67(4), 1174–1185.

Grubesic, T. H. (2012). The U.S. National Broadband Map: Data Limitations and
Implications. *Telecommunications Policy*, 36(2), 113–126. https://doi.org/https://doi.
org/10.1016/j.telpol.2011.12.006

Grubesic, T. H., & Matisziw, T. C. (2006). On the Use of ZIP Codes and ZIP Code
Tabulation Areas (ZCTAs) for the Spatial Analysis of Epidemiological Data.
International Journal of Health Geographics, 5(1), 58.

Hampton, K., Fernandez, L., Robertson, C., & Bauer, J. M. (2020). *Broadband and Student Performance Gaps*. James H. and Mary B. Quello Center, Michigan State University. https://doi.org/10.2139/ssrn.3614074

Hargittai, E. (2002). Second-Level Digital Divide: Differences in People's Online Skills. *First Monday*, *7*(4), 1–20.

Hargittai, E., & Hinnant, A. (2008). Digital Inequality: Differences in Young Adults' Use of the Internet. *Communication Research*, *35*(5), 602–621.

Hassani, S. N. (2006). Locating Digital Divides at Home, Work, and Everywhere Else. *Poetics*, *34*, 250–272.

Hauge, J. A., & Prieger, J. E. (2010). Demand-Side Programs to Stimulate Adoption of Broadband: What Works? *Review of Network Economics*, *9*(3), Article 4. https://doi.org/10.2202/1446-9022.1234

Hauge, J. A., & Prieger, J. E. (2015). Evaluating the Impact of the American Recovery and Reinvestment Act's BTOP on Broadband Adoption. *Applied Economics*, *47*(60), 6553–6579. https://doi.org/10.1080/00036846.2015.1080810

Helsper, E. J. (2012). A Corresponding Fields Model for the Links between Social and Digital Exclusion. *Communication Theory: CT: A Journal of the International Communication Association*, *22*(4), 403–426.

Helsper, E. J., & Eynon, R. (2013). Distinct Skill Pathways to Digital Engagement. *European Journal of Communication*, *28*(6), 696–713.

Helsper, E. J., Van Deursen, A. J. A., & Eynon, R. (2015). *Tangible Outcomes of Internet Use: From Digital Skills to Tangible Outcomes Project Report*. Oxford Internet Institute, University of Twente, and London School of Economics and Political Science. https://www.oii.ox.ac.uk/archive/downloads/publications/Tangible_Outcomes_of_Internet_Use.pdf

Hero, R. E. (2000). *Faces of Inequality: Social Diversity in American Politics*. Oxford University Press.

Hindman, M. (2008). *The Myth of Digital Democracy*. Princeton University Press.

Hinton, R. (2020). Cook County COVID-19 Response Plan Focuses on Communities "Hit the Hardest" to Ensure "Recovery Will Encompass Everyone." *Chicago Sun-Times*. https://chicago.suntimes.com/politics/2020/5/14/21258932/cook-county-covid-19-response-plan-digital-divide-transit-coronavirus-preckwinkle

Holloway, S. R., & Mulherin, S. (2004). The Effect of Adolescent Neighborhood Poverty on Adult Employment. *Journal of Urban Affairs*, *26*(4), 427–454.

Holmes, A., Bell Fox, E., Wieder, B., & Zuback-Skees, C. (2016). *Rich People Have Access to High-Speed Internet; Many Poor People Still Don't*. https://publicintegrity.org/business/rich-people-have-access-to-high-speed-internet-many-poor-people-still-dont/

Holt, L., & Jamison, M. (2009). Broadband and Contributions to Economic Growth: Lessons from the US Experience. *Telecommunications Policy*, *33*(10), 575–581.

Horrigan, J. B. (2015). The Training Difference: How Formal Training on the Internet Impacts New Users. *TPRC 43: The 43rd Research Conference on Communication, Information and Internet Policy*. https://doi.org/10.2139/ssrn.2587783

Horrigan, J. B. (2018). *Digital Skills and Job Training: Community-Driven Initiatives Are Leading the Way in Preparing Americans for Today's Jobs*. https://www.benton.org/publications/digital-skills-and-job-training-community-driven-initiatives-are-leading-way-preparing

Horrigan, J. B. (2019). *Smart Cities and Digital Equity*. https://www.digitalinclusion.org/wp-content/uploads/2019/06/NDIA_Horrigan-Smart-Cities_Final.pdf

Howard, B., & Morris, T. (2019). *Tribal Technology Assessment*. https://aipi.asu.edu/sites/default/files/tribal_tech_assessment_compressed.pdf

Ioannides, Y. M., & Datcher Loury, L. (2004). Job Information Networks, Neighborhood Effects, and Inequality. *Journal of Economic Literature, 42*(4), 1056–1093.

Jargowsky, P. A. (1997). *Poverty and Place*. Russell Sage Foundation.

Jayakar, K., & Park, E. A. (2013). Broadband Availability and Employment: An Analysis of County-Level Data from the National Broadband Map. *Journal of Information Policy, 3*, 181–200.

Jensen, N. (2019). Five Economic Development Takeaways from the Amazon HQ Bids. Brookings Institution, March 4. https://www.brookings.edu/research/five-economic-development-takeaways-from-the-amazon-hq2-bids/

Johnston, K. (2016). Google Helps Boston Nonprofit Expand Internet Training for Low-Income Residents. *Boston Globe*, February 17. https://www.bostonglobe.com/business/2016/02/17/boston-uses-google-grant-expand-low-income-tech-training/IOD2I8BIQYF0acNOtEFEdJ/story.html

Johnston, R. (2020a). Pandemic Gave Government's Digital Services a New "Baseline." *StateScoop*, October 2. https://statescoop.com/pandemic-government-digital-services-nikhil-deshpande/

Johnston, R. (2020b). Virtual City Council Meetings Are Boom or Bust for Local Governments. *StateScoop*, April 2. https://statescoop.com/virtual-city-council-meetings-boom-bust-local-government/

Jones, B. D., & Bachelor, L. W. (1993). *The Sustaining Hand: Community Leadership and Corporate Power?* University Press of Kansas.

Kaplan, D., & Mossberger, K. (2012). Prospects for Poor Neighborhoods in the Broadband Era: Neighborhood-Level Influences on Technology Use at Work. *Economic Development Quarterly, 26*(1), 95–105.

Karch, A., Nicholson-Crotty, S., Woods, N. D., & Bowman, A. O. (2016). Policy Diffusion and the Pro-Innovation Bias. *Political Research Quarterly, 69*(1), 83–95.

Karsten, J. (2019). New Training Models and Policies for a Digital Economy Workforce. Brookings Institution. https://www.brookings.edu/blog/techtank/2019/04/18/new-training-models-and-policies-for-a-digital-economy-workforce/

Katz, B., & Bradley, J. (2013). *The Metropolitan Revolution: How Cities and Metros Are Fixing Our Broken Politics and Fragile Economy*. Brookings Institution Press.

Katz, B., & Nowak, J. (2017). *The New Localism: How Cities Can Thrive in the Age of Populism*. Brookings Institution Press.

Kelly, M. (2020). *The FCC Has Received Hundreds of Complaints about Carriers' Coronavirus Pledge*. https://www.theverge.com/2020/5/19/21263843/fcc-ajit-pai-coronavirus-pandemic-keep-americans-connected-verizon-att-comcast

Kettl, D. F. (2020). States Divided: The Implications of American Federalism for Covid-19. *Public Administration Review, 80*(4), 596–602.

Kingdon, J. W. (1995). *Agendas Alternatives and Public Policies*. Longman Classics in Political Science.

Kneebone, E., & Holmes, N. (2016). *U.S. Concentrated Poverty in the Wake of the Great Recession*. https://www.brookings.edu/research/u-s-concentrated-poverty-in-the-wake-of-the-great-recession/

Koetsier, J. (2020). COVID-19 Accelerated E-Commerce Growth "4 to 6 Years." *Forbes Magazine*. https://www.forbes.com/sites/johnkoetsier/2020/06/12/covid-19-accelerated-e-commerce-growth-4-to-6-years/?sh=5acae20f600f

Koksal, I. (2020). The Rise of Online Learning. *Forbes*, May 2. https://www.forbes.com/sites/ilkerkoksal/2020/05/02/the-rise-of-online-learning

Kolko, J. (2010). *Does Broadband Boost Local Economic Development?* Public Policy Institute of California.

Kolko, J. (2012). Broadband and Local Growth. *Journal of Urban Economics*, *71*(1), 100–113.

Krause, E., & Sawhill, I. V. (2018). *Seven Reasons to Worry about the American Middle Class*. https://www.brookings.edu/blog/social-mobility-memos/2018/06/05/seven-reasons-to-worry-about-the-american-middle-class/

Kreitzer, R. J. (2015). Politics and Morality in State Abortion Policy. *State Politics & Policy Quarterly*, *15*(1), 41–66. https://doi.org/10.1177/1532440014561868

Kreitzer, R. J., & Boehmke, F. J. (2016). Modeling Heterogeneity in Pooled Event History Analysis. *State Politics & Policy Quarterly*, *16*(1), 121–141.

Laar, E. van, van Deursen, A. J. A. M., van Dijk, J. A. G. M., & de Haan, J. (2019). Determinants of 21st-Century Digital Skills: A Large-Scale Survey among Working Professionals. *Computers in Human Behavior, 100*, 93–104. https://doi.org/https://doi.org/10.1016/j.chb.2019.06.017

LaRose, G. (2018). *New Orleans Mayor LaToya Cantrell Launches Digital Equity Initiative: What Is It?* Nola.com.

LaRose, R., Gregg, J. L., Strover, S., Straubhaar, J., & Carpenter, S. (2007). Closing the Rural Broadband Gap: Promoting Adoption of the Internet in Rural America. *Telecommunications Policy*, *31*(6), 359–373.

LaRose, R., Strover, S., Gregg, J. L., & Straubhaar, J. (2011). The Impact of Rural Broadband Development: Lessons from a Natural Field Experiment. *Government Information Quarterly*, *28*(1), 91–100. https://doi.org/10.1016/j.giq.2009.12.013

Lax, J. R., & Phillips, J. H. (2009a). Gay Rights in the States: Public Opinion and Policy Responsiveness. *American Political Science Review*, *103*(3), 367–386.

Lax, J. R., & Phillips, J. H. (2009b). How Should We Estimate Public Opinion in the States? *American Journal of Political Science*, *53*(1), 107–121.

Layton, R. (2018). Where Are the Silicon Valley Firms When It Comes to Digital Inclusion in San Jose? *Forbes*. https://www.forbes.com/sites/roslynlayton/2018/11/16/where-are-the-silicon-valley-firms-when-it-comes-to-digital-inclusion-in-san-jose/#4ed94126115c

Lee, W. (2018). The Scramble for Sunnyvale: Tech Companies in "Arms Race" for Office Space. *San Francisco Chronicle*. https://www.sfchronicle.com/business/article/The-scramble-for-Sunnyvale-Tech-companies-in-12908684.php

Lehr, W., Sirbu, M., & Gillett, S. (2006). Wireless Is Changing the Policy Calculus for Municipal Broadband. *Government Information Quarterly, 23*(3), 435–453.

Leonhardt, D., & Serkez, Y. (2020). America Will Struggle after Coronavirus. *New York Times*, April 10. https://www.nytimes.com/interactive/2020/04/10/opinion/coronavirus-us-economy-inequality.html

Lewis, J. H., & Hamilton, D. K. (2011). Race and Regionalism: The Structure of Local Government and Racial Disparity. *Urban Affairs Review*, *47*(3), 349–384. https://doi.org/10.1177/1078087410391751

Li, A., & Sussman, J. (2018). *Bridging the Digital Divide.* Wharton Public Policy Initiative. https://publicpolicy.wharton.upenn.edu/live/news/2420-bridging-the-digital-divide

LISC. (2019). NACo, Rural LISC and RCAP Launch Mobile App and Announce the Bridging the Economic Divide Partnership to Address Rural Broadband Access. https://www.lisc.org/rural/regional-stories/naco-rural-lisc-and-rcap-launch-mobile-app-and-announce-bridging-economic-divide-partnership-address-rural-broadband-access/

LISC Chicago. (2009). *A Platform for Participation and Innovation.* http://archive.lisc-chicago.org/uploads/lisc-chicago/documents/scpmasterplan.pdf

Liu, A. (2019). A Better Way to Attract Amazon's Jobs. *New York Times.* https://www.nytimes.com/2019/02/16/opinion/amazon-new-york.html

Logan, J. R., & Stults, B. J. (2011). *The Persistence of Segregation in the Metropolis: New Findings from the 2010 Census.* https://s4.ad.brown.edu/Projects/Diversity/Data/Report/report2.pdf

Lohr, S. (2018). Digital Divide Is Wider Than We Think, Study Says. *New York Times,* December 4. https://www.nytimes.com/2018/12/04/technology/digital-divide-us-fcc-microsoft.html

Lohr, S. (2020). As Jobs Move Online, Retraining Workers Becomes a Priority. *New York Times.* https://www.nytimes.com/2020/07/13/business/as-jobs-move-online-retraining-workers-becomes-a-priority.html

Lucas, R. E. (1988). On the Mechanics of Economic Development. *Journal of Monetary Economics, 22,* 3–42.

Lyu, W., & Wehby, G. L. (2020). Community Use of Face Masks and COVID-19: Evidence from a Natural Experiment of State Mandates in the US: Study Examines Impact on COVID-19 Growth Rates Associated with State Government Mandates Requiring Face Mask Use in Public. *Health Affairs, 39*(8), 1419–1425.

Mack, E. A. (2014). Broadband and Knowledge Intensive Firm Clusters: Essential Link or Auxiliary Connection? *Papers in Regional Science, 93*(1), 3–29.

Mack, E. A., & Faggian, A. (2013). Productivity and Broadband: The Human Factor. *International Regional Science Review, 36*(3), 392–423.

Maggetti, M., & Gilardi, F. (2016). Problems (and Solutions) in the Measurement of Policy Diffusion Mechanisms. *Journal of Public Policy, 36,* 87–107. doi:10.1017/S0143814X1400035X

Mallinson, D. J. (2021). Who Are Your Neighbors? The Role of Ideology and Decline of Geographic Proximity in the Diffusion of Policy Innovations. *Policy Studies Journal 49*(1): 6–88.

Mallinson, D. J. (2020). The Spread of Policy Diffusion Studies: A Systematic Review and Meta-Analysis, 1990–2018. APSA Preprints. doi:10.33774/apsa-2020-csnt6.

Manlove, J., & Whitacre, B. (2019). An Evaluation of the Connected Nation Broadband Adoption Program. *Telecommunications Policy, 43*(7), 1.

Manta. (2019). Sunnyvale, IT Companies. https://www.manta.com/mb_53_G4_280/information_technology/sunnyvale_ca

Markle. (2019). *Digital Blindspot: How Digital Literacy Can Create a More Resilient American Workforce.* https://www.markle.org/sites/default/files/2019-10-24-RABN-Digital-Literacy-ReportFINAL.pdf

Markle. (2020). *Stimulus for American Opportunity.* https://www.markle.org/sites/default/files/Stimulus-For-American-Opportunity.pdf

Massey, D. S., & Denton, N. A. (1993). *American Apartheid: Segregation and the Making of the Underclass*. Harvard University Press.

Mayor's Advisory Council on Closing the Digital Divide. (2007). *The City That NetWorks: Transforming Society and Economy through Digital Excellence*. http://rw-ventures.com/wp-content/uploads/2017/01/DIGITAL-DIVIDE_web-version-0707.pdf

McAuley, J., & Leskovec, J. (2012). Learning to Discover Social Circles in Ego Networks. *Proceedings of the 25th International Conference on Neural Information Processing Systems*. http://i.stanford.edu/~julian/pdfs/nips2012.pdf

McCabe, D. (2020). Poor Americans Face Hurdles in Getting Promised Internet. *New York Times*, May 20. https://www.nytimes.com/2020/05/20/technology/coronavirus-broadband-discounts.html

McIlwain, C. D. (2020). *Black Software: The Internet & Racial Justice, from the AfroNet to Black Lives Matter*. Illustrated ed. Oxford University Press.

Michener, J. (2018). *Fragmented Democracy: Medicaid, Federalism, and Unequal Politics*. Cambridge University Press.

Microsoft. (2017). *A Rural Broadband Strategy: Connecting Rural America to New Opportunities*.

Milken Institute. (2018). Best-Performing Cities 2018: Where America's Jobs Are Created. https://milkeninstitute.org/reports/best-performing-cities-2018-where-americas-jobs-are-created

Mintrom, M. (2000). *Policy Entrepreneurs and School Choice*. Georgetown University Press.

Mintrom, M., & Vergari, S. (1998). Policy Networks and Innovation Diffusion: The Case of State Education Reform. *Journal of Politics*, 60(1), 126–148.

Moretti, E. (2012). *The New Geography of Jobs*. First Mariner Books.

Morris, T., Mossberger, K., & Parkhurst, N. D. (Forthcoming). Digital Governance in Indian Country. In E. Welch (Ed.), *E-Government Research Handbook*. Edward Elgar.

Mossberger, K. (2000). *The Politics of Ideas and the Spread of Enterprise Zones*. Georgetown University Press.

Mossberger, K., Benoit Bryan, J., & Brown, A. (2014). *Smart Communities Evaluation: Civic 2.0 Participant Surveys and Interviews with Partner Organizations*. https://techdatasociety.asu.edu/sites/default/files/uploads/smartcommunities_civic2surveyinterviews.pdf

Mossberger, K., Feeney, M., & Li, M.-H. (2014). *Smart Communities Evaluation: FamilyNet Centers*. https://techdatasociety.asu.edu/sites/default/files/uploads/smartcommunities_familynetcenters.pdf

Mossberger, K., Kaplan, D., & Gilbert, M. (2008). Going Online without Easy Access: A Tale of Three Cities. *Journal of Urban Affairs*, 30(5), 469–488.

Mossberger, K., & Tolbert, C. J. (2009). *Digital Excellence in Chicago: A Citywide View of Technology Use*. https://techdatasociety.asu.edu/sites/default/files/digital_excellence_study_2009.pdf

Mossberger, K., Tolbert, C. J., & Anderson, C. (2012). *Measuring Change in Internet Use and Broadband Adoption: Comparing BTOP Smart Communities and Other Chicago Neighborhoods*. http://www.broadbandillinois.org/uploads/cms/documents/chicagosmartcommunitiespcireport4.pdf

Mossberger, K., Tolbert, C. J., & Anderson, C. (2014). *Measuring Change in Internet Use and Broadband Adoption: Comparing BTOP Smart Communities and Other Chicago*

Neighborhoods [updated 2014]. April. https://techdatasociety.asu.edu/sites/default/files/uploads/smartcommunities_measuringinternetchangeinchicago.pdf

Mossberger, K., Tolbert, C. J., & Anderson, C. (2017). The Mobile Internet and Digital Citizenship in African-American and Latino Communities. *Information, Communication and Society, 20*(10), 1587–1606.

Mossberger, K., Tolbert, C. J., Bowen, D., & Jimenez, B. (2012). Unraveling Different Barriers to Internet Use: Urban Residents and Neighborhood Effects. *Urban Affairs Review, 48*(6), 771–810.

Mossberger, K., Tolbert, C. J., & Franko, W. (2013). *Digital Cities: The Internet and the Geography of Opportunity*. Oxford University Press.

Mossberger, K., Tolbert, C. J., & Gilbert, M. (2006). Race, Place, and Information Technology. *Urban Affairs Review, 41*(5), 583–620.

Mossberger, K., Tolbert, C. J., & LaCombe, S. (2020). *A New Measure of Digital Participation and Its Impact on Economic Opportunity*. https://techdatasociety.asu.edu/sites/default/files/white_paper_vf-final-march2020.pdf

Mossberger, K., Tolbert, C. J., & McNeal, R. (2008). *Digital Citizenship: The Internet, Society and Participation*. MIT Press.

Mossberger, K., Tolbert, C. J., & Stansbury, M. (2003). *Virtual Inequality: Beyond the Digital Divide*. Georgetown University Press.

Muro, M. (2020). Will the Covid-19 Pandemic Accelerate Automation? *The Economist*, April 22. https://eiuperspectives.economist.com/technology-innovation/will-covid-19-pandemic-accelerate-automation

NACo. (2019). *Counties Matter*. https://www.naco.org/sites/default/files/documents/CM_2019.pdf

Nam, T., & Pardo, T. A. (2011). Conceptualizing Smart City with Dimensions of Technology, People, and Institutions. *Proceedings of the 12th Annual International Conference on Digital Government Research*, 282–291. https://smartcitiescouncil.com/sites/default/files/public_resources/Conceptualizing smart city.pdf

National Congress of American Indians. (2020). *Tribal Nations & The United States: An Introduction*. https://ncai.org/about-tribes

National Digital Inclusion Alliance. (2020). 2020 Trailblazers. https://www.digitalinclusion.org/digital-inclusion-trailblazers/

National Skills Coalition. (2020). *Applying a Racial Equity Lens to Digital Literacy*. https://www.nationalskillscoalition.org/resources/publications/file/Digital-Skills-Racial-Equity-Final.pdf

Nicholson-Crotty, S., & Carley, S. (2018). Information Exchange and Policy Adoption Decisions in the Context of US State Energy Policy. *State Politics & Policy Quarterly, 18*(2), 122–147.

NLC. (2016). Cities 101—Consolidations. https://www.nlc.org/resource/cities-101-consolidations

NTIA. (1998). *Falling through the Net II: New Data on the Digital Divide*. https://www.ntia.doc.gov/legacy/ntiahome/net2/falling.html

NTIA. (2000). *Falling through the Net: Toward Digital Inclusion*. https://www.ntia.doc.gov/legacy/ntiahome/fttn00/front00.htm#Mineta

NTIA Data Central. (2020a). Data Explorer. Wired High-Speed Internet Used at Home, US. https://www.ntia.doc.gov/data/digital-nation-data-explorer#sel=wiredHighSpeedAtHome&demo=&pc=prop&disp=chart

NTIA. (2020b). Working with States to Solve the Broadband Challenge. https://www.ntia.
gov/blog/2020/working-states-solve-broadband-challenge

NTIA & NSF. (2017). *The National Broadband Research Agenda*. https://www.ntia.doc.
gov/files/ntia/publications/nationalbroadbandresearchagenda-jan2017.pdf

Nyczepir, D. (2018). Connecting People without Internet Access in Silicon Valley. *Route
Fifty—State and Local News*, November 8. https://www.routefifty.com/tech-data/2018/
11/digital-divide-san-jose/152698/

OECD. (2019). OECD Broadband Portal. https://www.oecd.org/sti/broadband/
broadband-statistics/

Orfield, G., & Lee, C. (2005). *Why Segregation Matters: Poverty and Educational Inequality*.
Civil Rights Project at Harvard University. https://files.eric.ed.gov/fulltext/ED489186.pdf

Orfield, G., Losen, D., Wald, J., & Swanson, C. B. (2004). *Losing Our Future: How Minority
Youth Are Being Left Behind by the Graduation Rate Crisis*. Civil Rights Project at
Harvard University. http://files.eric.ed.gov/fulltext/ED489177.pdf

Owens, A. (2016). Inequality in Children's Contexts: Income Segregation of Households
with and without Children. *American Sociological Review*, 81(3), 549–574.

Pacheco, I., & Ramachandran, S. (2019). Do You Pay Too Much for Internet Service? See
How Your Bill Compares. *WSJ Online*. https://www.wsj.com/articles/do-you-pay-too-
much-for-internet-service-see-how-your-bill-compares-11577199600

Pacheco, J. (2011). Using National Surveys to Measure Dynamic U.S. State Public
Opinion: A Guideline for Scholars and an Application. *State Politics & Policy Quarterly*,
11(4), 415–439.

Pacheco, J. (2012). The Social Contagion Model: Exploring the Role of Public Opinion
on the Diffusion of Antismoking Legislation across the American States. *Journal of
Politics*, 74(1), 187–202.

Pacific Market Research. (2019). *Seattle IT Connectedness Segmentation Study*.
http://www.seattle.gov/Documents/Departments/SeattleIT/DigitalEngagement/
TechAccess/CityofSeattle Full Technical Report FINAL 01152019.pdf

Painter, M., & Qiu, T. (2020). *Political Beliefs Affect Compliance with COVID-19 Social
Distancing Orders*. https://doi.org/10.2139/ssrn.3569098

Parilla, J. (2017). *Opportunity for Growth*. Brookings Institution. https://www.brookings.
edu/wp-content/uploads/2017/09/metro_20170927_opportunity-for-growth-iedl-
report-parilla-final.pdf.

Parilla, J., Liu, S., & Whitehead, B. (2020). *How Local Leaders Can Stave Off a Small
Business Collapse from COVID-19*. Brookings Institution. https://www.brookings.edu/
research/how-local-leaders-can-stave-off-a-small-business-collapse-from-covid-19/

Park, D. K., Gelman, A., & Bafumi, J. (2004). Bayesian Multilevel Estimation with
Poststratification: State-Level Estimates from National Polls. *Political Analysis*, 12, 375–
385. http://www.stat.columbia.edu/~gelman/research/published/parkgelmanbafumi.pdf

Park, D. K., Gelman, A., & Bafumi, J. (2006). State-Level Opinions from National
Surveys: Poststratification Using Multilevel Logistic Regression. In Cohen, J. E. (Ed.),
Public Opinion in State Politics (pp. 209–229). Stanford University Press.

Park, S. (2017). *Digital Capital*. Palgrave Macmillan UK.

Parker, K., Horowits, J. M., Brown, A., Fry, R., Cohn, D., & Igielnik, R. (2018). Demographic
and Economic Trends in Urban, Suburban and Rural Communities. https://
www.pewsocialtrends.org/2018/05/22/demographic-and-economic-trends-in-
urban-suburban-and-rural-communities/

Pew Charitable Trusts. (2019). How State Policy Shapes Broadband Deployment. https://www.pewtrusts.org/en/research-and-analysis/issue-briefs/2019/12/how-state-policy-shapes-broadband-deployment

Pew Research Center. (2019). Demographics of Internet and Home Broadband Usage in the United States. Pew Research Center: Internet, Science & Tech. https://www.pewresearch.org/internet/fact-sheet/internet-broadband/

Philpott, C., Parker, J., & Wyatt, T. (2020). Sudden Loss of Smell—Why It Is a Reason to Self-Isolate. *The Conversation*. http://theconversation.com/sudden-loss-of-smell-why-it-is-a-reason-to-self-isolate-138682

Prieger, J. E. (2013). The Broadband Digital Divide and the Economic Benefits of Mobile Broadband for Rural Areas. *Telecommunications Policy*, *37*(6), 483–502.

Prior, M. (2006). The Incumbent in the Living Room: The Rise of Television and the Incumbency Advantage in US House Elections. *Journal of Politics*, *68*(3), 657–673.

Quaintance, Z. (2020). Covid-19 Pushes Digital Services from Luxury to Necessity. *Government Technology*, December. https://www.govtech.com/civic/COVID-19-Pushes-Digital-Services-from-Luxury-to-Necessity.html

Ragnedda, M. (2019). Conceptualizing Digital Capital. *Telematics and Informatics*, *35*(8), 2366–2375.

Raudenbush, S. W., & Bryk, A. S. (2002). *Hierarchical Linear Models: Applications and Data Analysis Methods (Advanced Quantitative Techniques in the Social Sciences)*. Sage Publications.

Reese, L. A., & Sands, G. (2017). Is Detroit Really Making A Comeback? *CityLab*. https://www.citylab.com/equity/2017/02/detroits-recovery-the-lass-is-half-full-at-most/517194/

Rhinesmith, C. (2016). *Digital Inclusion and Meaningful Broadband Adoption Initiatives*. Benton Foundation. https://www.benton.org/sites/default/files/broadbandinclusion.pdf

Romm, T. (2020). "It Shouldn't Take a Pandemic": Coronavirus Exposes Internet Inequality among US Students as Schools Close their Doors. *Washington Post*, March 16. https://www.washingtonpost.com/technology/2020/03/16/schools-internet-inequality-coronavirus/

Rose, S. J. (2018). How Different Studies Measure Income Inequality in the US. Urban Institute. https://www.urban.org/sites/default/files/publication/99455/how_different_studies_measure_income_inequality_0.pdf

Rosebud County. (n.d.a). History of Rosebud County. https://rosebudcountymt.gov/resources/

Rosebud County. (n.d.b). Home. Welcome to Rosebud County. https://rosebudcountymt.gov/

Saez, E., & Zucman, G. (2014). *Wealth Inequality in the United States since 1913: Evidence from Capitalized Income Tax Data* (No. 20625). https://www.nber.org/papers/w20625

Salemink, K., Strijker, D., & Bosworth, G. (2017). Rural Development in the Digital Age: A Systematic Literature Review on Unequal ICT Availability, Adoption, and Use in Rural Areas. *Journal of Rural Studies*, *54*, 360–371.

San Jose. (n.d.). Digital Inclusion and Broadband Strategy. https://www.sanjoseca.gov/your-government/departments-offices/office-of-the-city-manager/civic-innovation/digital-inclusion-and-broadband-strategy

San Jose Digital Inclusion Fund. (n.d.) Digital Inclusion by the numbers. . . https://www.sjdigitalinclusion.org/

Saunders, P. (2018). Detroit, Five Years after Bankruptcy. *Forbes*. https://www.forbes.com/sites/petesaunders1/2018/07/19/detroit-five-years-after-bankruptcy/#285bb524cfeb

Savage, R. L. (1985). Diffusion Research Traditions and the Spread of Policy Innovations in a Federal System. *Publius: The Journal of Federalism, 15*(4), 1–28. https://doi.org/10.1093/oxfordjournals.pubjof.a037561

Savitch, H. V., & Adhikari, S. (2017). Fragmented Regionalism: Why Metropolitan America Continues to Splinter. *Urban Affairs Review, 53*(2), 381–402.

Sawhill, I. V. (2018). *The Forgotten Americans: An Economic Agenda for a Divided Nation*. Yale University Press.

Schartman-Cycyk, S., Mossberger, K., Callahan, B., Novak, S., Sheon, A., Siefer, A., . . . Cho, S. K. (2019). *Connecting Cuyahoga Investment in Digital Inclusion Brings Big Returns for Residents and Administration*. https://static1.squarespace.com/static/59d3bca38dd041c401d9ed80/t/5d5c448f6fede80001a75334/1566328034944/Connecting+Cuyahoga_2019.pdf

Scheerder, A., van Deursen, A., & van Dijk, J. (2017). Determinants of Internet Skills, Uses and Outcomes. A Systematic Review of the Second- and Third-Level Digital Divide. *Telematics and Informatics, 34*(8), 1607–1624.

Schradie, J. (2011). The Digital Production Gap: The Digital Divide and Web 2.0 Collide. *Poetics, 39*(2), 145–168.

Shearer, R., Friedhoff, A., Shah, I., & Berube, A. (2018). *Metro Monitor 2018*. https://www.brookings.edu/wp-content/uploads/2018/02/2018-02_brookings-metro_metro-monitor-2018__final.pdf

Shearer, R., & Shah, I. (2018). Opportunity Industries Exploring the Industries That Concentrate Good and Promising Jobs in Metropolitan America. Brookings Institution. https://www.brookings.edu/wp-content/uploads/2018/12/2018.12_BrookingsMetro_Opportunity-Industries_Report_Shearer-Shah.pdf

Sheon, A. R., & Carroll, L. (2019). How Can Health Systems Leverage Technology to Engage Patients? In E. W. Marx (Ed.), *Voices of Innovation: Fulfilling the Promise of Information Technology in Healthcare*. CRC Press.

Shipan, C. R., & Volden, C. (2006). Bottom-Up Federalism: The Diffusion of Antismoking Policies from US Cities to States. *American Journal of Political Science, 50*(4), 825–843.

Shipan, C. R., & Volden, C. (2008). The Mechanisms of Policy Diffusion. *American Journal of Political Science, 52*(4), 840–857.

Siefer, A. (2016). *Boston's Broadband and Digital Equity Advocate*. National Digital Inclusion Alliance. https://www.digitalinclusion.org/blog/2016/02/14/bostons-broadband-and-digital-equity-advocate

Siefer, A. (2017). *FCC Suggested Polices to "Bridge the Digital Divide" Do No Such Thing*. National Digital Inclusion Alliance. https://www.digitalinclusion.org/blog/2017/11/13/fcc-not-bridging-digital-divide/

Siefer, A., & Callahan, B. (2020). *Limiting Broadband Investment to "Rural Only" Discriminates against Black Americans and Other Communities of Color*. National Digital Inclusion Alliance. https://www.digitalinclusion.org/digital-divide-and-systemic-racism/

Simama, J. (2020). "It's 2020. Why Is the Digital Divide Still with Us?" *Governing*. https://www.governing.com/now/Its-2020-Why-Is-the-Digital-Divide-Still-with-Us.html

Sisson, P. (2018). *Why "Micropolitan" Cities May Be the Key to Rural Resurgence*. Curbed. https://archive.curbed.com/2018/10/30/18042760/job-business-rural-economic-micropolitan

Smart Cities Council. (2016). *Smart Infrastructure Unlocks Equity and Prosperity for Our Cities and Towns*. https://smartcitiescouncil.com/system/tdf/main/public_resources/SCC_Policy%20Brief_Smart%20Infrastructure_Exec_Summary.pdf?file=1&type=node&id=4168&force=

Smith, A., & Page, D. (2015). *U.S. Smartphone Use in 2015*. https://www.pewresearch.org/wp-content/uploads/sites/9/2015/03/PI_Smartphones_0401151.pdf

Smith, B. (2020). *Microsoft Launches Initiative to Help 25 Million People Worldwide Acquire the Digital Skills Needed in a COVID-19 Economy*. Microsoft. https://blogs.microsoft.com/blog/2020/06/30/microsoft-launches-initiative-to-help-25-million-people-worldwide-acquire-the-digital-skills-needed-in-a-covid-19-economy/

Snijders, T. A., & Bosker, R. J. (2012). *Multilevel Analysis: An Introduction to Basic and Advanced Multilevel Modeling*. Sage Publications.

Snyder, J. M., Jr., & Strömberg, D. (2008). *Press Coverage and Political Accountability* (No. w13878). National Bureau of Economic Research. https://doi.org/10.3386/w13878

Sovey, A. J., & Green, D. P. (2011). Instrumental Variables Estimation in Political Science: A Readers' Guide. *American Journal of Political Science*, 55(1), 188–200.

Speed Up San Jose. (n.d.). https://www.sanjoseca.gov/your-government/departments/office-of-the-city-manager/offices/civic-innovation/projects/speedup-san-jose

Stauffer, A., de Wit, K., Read, A., & Kitson, D. (2020). *How States Are Expanding Broadband Access*. Pew Charitable Trusts. https://www.pewtrusts.org/en/research-and-analysis/reports/2020/02/how-states-are-expanding-broadband-access

Steenbergen, M. R., & Jones, B. S. (2002). Modeling Multilevel Data Structures. *American Journal of Political Science*, 46(1), 218–237.

Stenberg, P., Morehart, M., Vogel, S., Cromartie, J., Breneman, V., & Brown, D. (2009). *Broadband Internet's Value for Rural America*. US Department of Agriculture Economic Research Service. https://www.ers.usda.gov/publications/pub-details/?pubid=46215

Stephens-Davidowitz, S. (2020). Google Searches Can Help Us Find Emerging Covid-19 Outbreaks. *New York Times*. https://www.nytimes.com/2020/04/05/opinion/coronavirus-google-searches.html

Stewart, N. (2020). She's 10, Homeless and Eager to Learn. But She Has No Internet. *New York Times*, March 26. https://www.nytimes.com/2020/03/26/nyregion/new-york-homeless-students-coronavirus.html

Streitfeld, D. (2018). Was Amazon's Headquarters Contest a Bait and Switch? Critics Say Yes. *New York Times*, November 6. https://www.nytimes.com/2018/11/06/technology/amazon-hq2-long-island-city-virginia.html

Strover, S., Whitacre, B., Rhinesmith, C., & Schrubbe, A. (2017). At the Edges of the National Digital Platform. *D-Lib Magazine*. http://www.dlib.org/dlib/may17/strover/05strover.html

Strover, S. (Forthcoming). Broadband for Telemedicine and Health Services. In K. Mossberger, E. Welch, & Y. Wu (Eds.), *Transforming Everything? Evaluating Broadband's Impacts across Policy Areas* (pp. 169–190). Oxford University Press.

Sunstein, C. R. (2007). *Republic.com 2.0*. Princeton University Press.

Tolbert, C. J., & Mossberger, K. (2015). *Final Report: Digital Inequality across Local Geographic Areas in the United States—States, Counties, Metro Areas and Principal Cities 1997–2014*. NSF BCC: Broadband Use Mapping, Data and Evaluation. NSF #1338471.

Tolbert, C. J., Mossberger, K., Gaydos, N., & Caldarulo, M. (Forthcoming). Addressing Spatial Inequality in Broadband Use and Community/Neighborhood Level Outcomes.

In E. Welch, Y. Wu, & K. Mossberger (Eds.), *Transforming Everything? Evaluating Broadband's Impacts across Policy Areas*. Oxford University Press.

Tomer, A., Kneebone, E., & Shivaram, R. (2017). *Signs of Digital Distress: Mapping Broadband Availability and Subscription in American Neighborhoods*. Brookings Institution. https://www.brookings.edu/wp-content/uploads/2017/09/broadbandreport_september2017.pdf

Tomer, A., Kneebone, E., & Shivaram, R. (2017). *Signs of Digital Distress: Mapping Broadband Availability and Subscription in American Neighborhoods*. https://www.brookings.edu/research/signs-of-digital-distress-mapping-broadband-availability/

Tomer, A., & Shivaram, R. (2017). Rollback of the FCC's Lifeline Program Can Hurt Households That Need Broadband the Most. *The Brookings Institution*, November. https://www.brookings.edu/blog/the-avenue/2017/11/27/rollback-of-the-fccs-lifeline-program-can-hurt-households-that-need-broadband-the-most/

Townsend, A. M. (2013). *Smart Cities: Big Data, Civic Hackers, and the Quest for a New Utopia*. W. W. Norton & Company.

Trounstine, J. (2018). *Segregation by Design: Local Politics and Inequality in American Cities*. Cambridge University Press.

Trussler, M. J. (2019). *The Impact of High Information Environments on Representation in the U.S. House of Representatives*. PhD Dissertation, Vanderbilt University. https://ir.vanderbilt.edu/bitstream/handle/1803/13910/TrusslerVanderbiltDissertationAug16.pdf?sequence=1

Turner Lee, N. (2019). *Enabling Opportunities: 5G, the Internet of Things, and Communities of Color*. https://www.brookings.edu/research/enabling-opportunities-5g-the-internet-of-things-and-communities-of-color/

Universal Service Administrative Company. (2020). *Lifeline Program*. https://www.usac.org/lifeline/

Ungerer, C., & Portugal, A. (2020). *Leveraging E-Commerce in the Fight against COVID-19*. Brookings Institution. https://www.brookings.edu/blog/future-development/2020/04/27/leveraging-e-commerce-in-the-fight-against-covid-19/

Urban, R., Bergen, M., & Bass, D. (2018). Google's Growing Empire May Transform Biggest City in Bay Area. *Bloomberg*. https://www.bloomberg.com/news/articles/2018-03-01/google-s-growing-empire-seen-transforming-biggest-bay-area-city

US Census Bureau. (2017a). *American Community Survey 1-year Public Use Microdata Samples*. Table S1901.

US Census Bureau. (2017b). *American Community Survey 1-year Public Use Microdata Samples*. Table S2801.

US Department of Agriculture. (2019). *A Case for Rural Broadband*. https://www.usda.gov/sites/default/files/documents/case-for-rural-broadband.pdf

US Ignite. (n.d.). Communities, Flint MI. Industrious Legacy Meets Forward-Thinking Connectivity. https://www.us-ignite.org/communities/flint-mi/

Van Deursen, A. J. A. M., & Helsper, E. J. (2015). The Third-Level Digital Divide: Who Benefits Most from Being Online? In L. Robinson, S. R. Cotten, J. Schulz, T. M. Hale, & A. Williams (Eds.), *Communication and Information Technologies Annual* (pp. 29–52). Emerald. https://doi.org/10.1108/S2050-206020150000010002

Van Deursen, A., & Van Dijk, J. (2011). Internet Skills and the Digital Divide. *New Media and Society*, 13(6), 893–911.

Van Deursen, A. J., & Van Dijk, J. A. (2014). The Digital Divide Shifts to Differences in Usage. *New Media & Society*, 16(3), 507–526.

Vigdor, J. L., & Ladd, H. F. (2010). *Scaling the Digital Divide: Home Computer Technology and Student Achievement* (No. w16078). National Bureau of Economic Research. https://doi.org/10.3386/w16078

Vogels, E., Perrin, A., Rainie, L., & Anderson, M. (2020). *53% of Americans Say the Internet Has Been Essential during the COVID-19 Outbreak.* Pew Research Center. https://www.pewresearch.org/internet/2020/04/30/53-of-americans-say-the-internet-has-been-essential-during-the-covid-19-outbreak/

Walker, J. L. (1969). The Diffusion of Innovations among the American States. *American Political Science Review, 63*(3), 880–899.

Wellenius, G. A., Vispute, S., Espinosa, V., Fabrikant, A., Tsai, T. C., Hennessy, J., . . . Gabrilovich, E. (2020). Impacts of US State-Level Social Distancing Policies on Population Mobility and COVID-19 Case Growth During the First Wave of the Pandemic. *arXiv.* http://arxiv.org/abs/2004.10172

West, D. M. (2018). *The Future of Work: Robotics, AI and Automation.* Brookings Institution Press.

Whitacre, B., & Brooks, L. (2014). Do Broadband Adoption Rates Impact a Community's Health? *Behaviour & Information Technology, 33*(7), 767–779.

Whitacre, B., & Gallardo, R. (2020). State Broadband Policy: Impacts on Availability. *Telecommunications Policy, 44*(9). https://doi.org/https://doi.org/10.1016/j.telpol.2020.102025.

Whitacre, B., Gallardo, R., & Strover, S. (2014a). Broadband's Contribution to Economic Growth in Rural Areas: Moving towards a Causal Relationship. *Telecommunications Policy, 38*(11), 1011–1023.

Whitacre, B., Gallardo, R., & Strover, S. (2014b). Does Rural Broadband Impact Jobs and Income? Evidence from Spatial and First-Differenced Regressions. *Annals of Regional Science, 53*(3), 649–670.

Whitacre, B., Strover, S., & Gallardo, R. (2015). How Much Does Broadband Infrastructure Matter? Decomposing the Metro–Non-Metro Adoption Gap with the Help of the National Broadband Map. *Government Information Quarterly, 32*(3), 261–269.

Wiggers, K. (2019). Comcast Expands Internet Essentials to Almost 3 Million More Low-Income Households. https://venturebeat.com/2019/08/06/comcast-expands-internet-essentials-to-almost-3-million-more-low-income-households/

Wilson, F., Brown, M., Cockhren, J., Williams, R., & Elliott, F. (2019). *Smart Black Tech Ecosystems. An In-Depth Analysis of Racial Equity in the Chicago Tech Sector.* Black Tech Mecca, June. https://www.blacktechmecca.org/research

Wilson, W. J. (1987). *The Truly Disadvantaged: The Inner City, the Underclass, and Public Policy.* University of Chicago Press.

Index

For the benefit of digital users, indexed terms that span two pages (e.g., 52–53) may, on occasion, appear on only one of those pages.

Tables and figures are indicated by *t* and *f* following the page number.